Greta's Homework

The Most Important Book You'll Ever Read: 101 Truths about Climate Change that Everyone Should Know (Especially Hysterical, Sanctimonious Hypocritical Mythmakers)

Zina Cohen

This short, fact-packed book utterly demolishes the climate change argument put forward by Greta Thunberg and the army of global warming hysterics who are attempting to dictate the way our world is run, and who are doing far more harm than good. The aim of this book is to rescue Ms Thunberg, and millions of others, from the slough of despond into which they have been propelled by pseudoscientific nonsenses and celebrity hysteria.

Copyright 2020 Zina Cohen
The right of Zina Cohen to be identified as the author of this work has been asserted in accordance with the Copyright, Designs and Patents Act 1988.

Published by BM Marvel Publishing, London

The author
Zina Cohen is the best-selling author of *The Shocking History of the EU*.

Preface

In 2019, Greta Thunberg, a young Swedish girl who is said to suffer from a cocktail of mental illnesses including Asperger's syndrome, obsessive-compulsive disorder and selective mutism, became a poster-girl for the climate change campaign.

Ms Thunberg often appears to be dogmatic, arrogant and rude and to lack humility and respect but it is not unreasonable to assume that these are a result of her collection of illnesses. She appears genuine and whole hearted, and I do not doubt for a moment that her concern for the environment and the planet is well-founded. She clearly believes what she says.

But she's wrong, of course.

The planet is not under threat. The climate change scare is nothing more than that – a scare.

But Ms Thunberg's enthusiasms have been so well publicised (and, dare I say, promoted) that there is no doubt in my mind that she has, quite unnecessarily, terrified millions of children around the world who now also believe that climate change threatens their future. Ms Thunberg, I fear, has done considerably more harm than good. By scaring a whole generation of children, I fear that young Ms Thunberg has created an epidemic of fear and anxiety.

More importantly, the ill-conceived changes which are being proposed by those who believe the pseudoscience known as global warming, will do a massive amount of very real harm.

And that leads to the big question: who is responsible for turning a young Swedish girl into a global icon, a Pied Piper, a 21st century Joan of Arc?

From a distance, it does seem to me that there may have been a good deal of management going on.

Who fed Greta Thunberg the information which has upset her so much? Why does she believe the nonsense that man-made climate change threatens our planet?

Was it Greta's decision alone to sail across the Atlantic when she wanted to visit America? Or did someone else suggest that it would be a good publicity stunt?

Sailing across the Atlantic didn't help the planet since at least two crew members are reported to have flown across the sea in order to ensure that Ms Thunberg wasn't seen boarding an aeroplane. You don't need to be good at maths to realise that the planet would have been better off if Ms Thunberg had forgotten about the boat and had taken a commercial flight to America. But was the boat really Ms Thunberg's idea? Did she sit up one morning and say: 'I think I will go to America in an expensive boat'?

There has not been so much disinformation and pseudoscience in evidence since the height of the AIDS nonsense at the end of the last century. And so I have compiled this short book to help balance the pseudoscience and the barrage of misinformation which has characterised the global warming scare.

I hope that Ms Thunberg, and those who share her distress and fear, many of whom are old enough to know better, will read this collection of facts and realise that they have been misled and manipulated.

The world will be a safer, happier place if those sanctimonious hypocrites who fly around the world promoting the nonsense known as climate change take note.

And when this fashionable scare story is over, maybe Ms Thunberg, and many other children, will be able to enjoy life a little more.

Zina Cohen

Foreword

I used to be a passionate believer in climate change.

I was writing about it when it was called global warming several decades ago, and being widely vilified for my fears – probably by the same people who are now enthusiastic 'believers'. In those days, people like me were worried about the ozone layer and greenhouse gases.

Even then the whole issue wasn't new.

Climate change was first written about by a Swedish scientist called Svante Arrhenius in 1896. He predicted that fossil fuel emissions might delay the onset of another ice age. I bet that most of those who are now most vocal in their insistence that the end of the world is nigh have never heard of Svante Arrhenius.

Today's campaigners (largely an unpleasant mixture of children and celebrities, seasoned with a few utterly insane eco-warriors) like to think they invented the idea of climate change but pseudo-scientists have, since the 19th century, been putting forward the theory that greenhouse gases could change the climate. Only the evangelical mythmakers, consumed by hubris, think the evidence has suddenly become convincing.

Today, only nutters and pseudoscientists who see the whole business as a way to become rich and famous think the evidence is convincing. Independent scientists admit that there is no genuine evidence that the earth is heating up.

Moreover, even if the earth is getting hotter, there is no way to know if this could be caused by burning fossil fuels. Maybe the earth's core is heating up? There is a nauseating, sanctimonious (and usually hypocritical) arrogance about those who believe that if the planet is changing then it must be because of Man (and Woman, of course.)

Those opposed to industry and capitalism have always blamed natural catastrophes on man's behaviour but a careful study of the real facts has convinced me that the problem has been wildly exaggerated. The problem is that the Climate Change Mythmakers (CCMs) exchange fake facts as though they were baseball cards, and the whole climate change fraud can be traced back to a now infamous DVD made by Al Gore. The BBC has kept the deceit going with (among others) a number of programmes made by David Attenborough which should have been categorised as drama rather than documentaries.

I have, during my life, studied many scientific, political and historical controversies and tried to analyse the truths and see through the lies, the deceptions and the politically convenient half-truths. Today I have absolutely no doubt that the climate change scare, and the argument that if the climate is changing it is a man-made phenomenon, can no more be substantiated than the widely exaggerated AIDS scare which was promoted with such enthusiasm in the 1990s.

The climate change myth makers are part of a huge and profitable industry, and most of those promoting the myth have a vested interest in its survival and its growth. There are books to be sold, television programmes to be made, conferences to organise and billions to be made out of building wind farms and solar farms. A lot of people have become very rich as a result of the climate change nonsense.

Despite the absence of any evidence, the climate change nonsense is widely supported by people who ought to know better but who want to clamber onto a fashionable bandwagon. The British Government's Department of Energy has even changed its name to the Department of Energy and Climate Change and has its new name expensively installed outside the front door.

The leading advocates of climate change include the quaintly potty and naïve meat eating Prince Charles, who now seems to have conveniently forgotten that a few years ago he forecast that the world would end in 2017, an assortment of slightly hysterical priggish children who are enjoying the freedom to join a campaign about which they know absolutely nothing, the usual crew of luvvies (most of whom also opposed the democratic vote to leave the fascist EU) and who are now flying around the world getting publicity by attending interminable climate change conferences, and what seems like the entire editorial staff of the once reliable BBC, who seem happy to give airtime to any minor celebrity or pseudo-scientist who is prepared to enthuse about the dangers of climate change while denying any airtime at all to scientists who bravely refuse to follow the official BBC line.

Climate change enthusiasts have promoted their cause by throwing bombs at policemen, by holding demonstrations designed to block the traffic and pollute the air and by forging documents purporting to give their spurious arguments some sort of a scientific basis.

The whole global warming fiasco is most notably marked by smug, self-righteous, ignorant celebrities who are delighted by their own certainty and the knowledge that whatever constraints are introduced should not apply to them. What, I wonder, are these celebrities and royals thinking when they fly by private plane to conferences where they pontificate on the subject of global warming (a subject about which most of them know bugger all)? Why do they think they deserve to be allowed to behave differently? Do they really consider themselves to be so special and superior? 'Do as I say, not what I do.' Are they stupid or mad? How stupid do they think we are?

(Greta Thunberg cannot escape this criticism. It is true that she travelled to America by boat. But at least two of the crew members had to fly across the Atlantic in order to look after the boat trip for her. Travelling by boat did nothing to prevent climate change. It is difficult to avoid the suspicion that the public show of self-righteousness may have been created for the sake of publicity rather than to protect the planet.)

I would be more impressed by their earnestness if the celebrities vowed to fly no more and to turn their aeroplanes into tractors. Those who are most vocal in warning us all that we must not fly, drive our cars or heat our homes are the ones who spend much of their lives in their aeroplanes or their expensive limousines; and when they aren't moving about the planet they are relaxing in one of their many luxurious, appropriately air-conditioned or well-heated homes. Prince Charles, for example, seems unable to pop to the shops without using a couple of private jets and a helicopter. Heavens, he probably uses a helicopter if he wants to go to the bathroom. He and his son Harry are typical examples of the modern, over-indulged hypocrite. Little Greta, the newly anointed replacement for Joan of Arc, travelled across the Atlantic in a boat which was allegedly equipped with two diesel engines. And there seems to have been a good deal of plastic involved too. A picture of her travelling by train showed her with a number of plastic food containers. (As someone remarked, celebrities must use a special type of plastic. The only harmful plastic is the sort we common people favour.)

Despite the hot air (not good for the planet) and the bleatings, the evidence supporting the climate change argument is thinner and weaker than a wet tissue. The global temperature (as measured by satellite) really hasn't risen (though climate change propagandists have tried to 'adjust' the figures to show that it has). The computer modelled predictions which are regularly used to frighten us all have been shown to be plain, old-fashioned wrong. Polar bear populations recently reached a 50 year high and the amount of ice in Antarctica is growing. These are facts which don't fit the hoax being perpetrated by the climate change conspiracy theorists but they are facts nevertheless.

And the perhaps rather surprising fact is that a recent International Panel on Climate Change published an Assessment Report which concluded that there really isn't any evidence supporting the theory that climate change has led to an increase in extreme weather events.

Nevertheless, there are now very serious attempts afoot to make it just as illegal even to question climate change as it is to question the holocaust.

Those who dare to question the idea of man-made climate change affecting our lives are dismissed as 'climate change deniers' in the same way that those who question the accepted truths about the holocaust are branded holocaust deniers. This seems to me to trivialise the holocaust in a most offensive way, and I am surprised that CCMs see nothing wrong with equating the pseudoscience of climate change with the holocaust.

We need a word to describe those idiots who believe the garbage produced by the climate change believers.

Hysterics, nutters or lunatics would seem appropriate but I prefer to refer to them as CCMs (Climate Change Mythmakers).

Climate Change Mythmakers are dim-witted, ill-informed, self-indulgent exhibitionists. They do not have the brains to realise that destroying public and private property leads only to a waste of energy and resources and that blocking traffic merely leads to frustration, stress and pollution. The climate change activists have done nothing whatsoever to improve the planet – on the contrary, their actions have done enormous and lasting damage.

It's all about 'look at me what a good person I am' – and let's have a smashing good time as well.

Greens are people who insist on blaming other people for their own failures and inadequacies and who wish others to suffer so that they can bask in the sunshine of their own self-righteousness.

In the early days there did seem to be a real chance that global warming would prove to be a huge problem. But the evidence supporting the theory has evaporated, if it ever truly existed, and today the argument for climate change is a spurious one, unsupported by serious science.

As I pointed out earlier, the climate change fraudsters now refer to those who don't accept their lies as 'deniers'. So now we have climate change deniers in the same way that there are holocaust deniers.

When you have read this book you will understand why I say that the world would be a happier, healthier, safer place if Greta stopped telling us all what to do, went back to school – and stayed there until she had learned a little more.

Zina Cohen
March 2020

1.

It was recently announced with much fanfare that Britain now obtains more electricity from renewables than from fossil fuels. But most of the electricity from 'renewables' comes from something called biomass. (Wind and solar power probably provide nearly enough electricity to power the computers and phones of climate change protestors.)

So, what is biomass?

I rather suspect that quite a lot of people don't know that 'biomass' is another word for 'wood' – the stuff you get from chopping down trees. (It is perfectly possible that some of the children who are hysterical about climate change, particularly those who have spent a good deal of time away from school, don't realise that wood comes from trees.)

And the trees which give us wood (aka biomass) are the same trees which we are told we must plant to save the planet.

The European Union and the British Government want to stop us using log burning stoves and open fireplaces. But at the same time they are shovelling tons of trees into power station burners.

This is more than lunacy. It's fraud.

The biomass enthusiasts are, presumably, the same idiots who thought biofuels were a good idea – not realising that burning food condemns millions to starve.

Still, it is, I suppose, now possible for the Climate Change Mythmakers (CCMs) to claim that electric cars are using electricity which grows on trees – rather than relying on the burning of diesel or coal.

2.

Journalists like to claim that every odd weather event is a result of climate change. This is just plain silly. It is, for example, alleged that the Victoria Falls haven't been as insipid as they are now for 25 years. This, we are told, is a result of climate change. So why were the Falls just as bad 25 years ago?

The facts show that nature is always changing and many of the things which have been observed in the last few years are neither new nor a reason to become hysterical.

There is a good deal of pseudoscience in many of the proclamations made about global warming.

3.

The global warming/global chilling/climate change myth is heavily promoted because thousands of people have a vested interest in its success. A lot of people have become very rich and famous as a result of the climate change nonsense.

Anyone who questions the theory is likely to be ostracised.

It is wise to remember Lenin and to look for who stands to gain.

So, who benefits from the climate change nonsense?

Well, for starters, there are some pretty huge industries involved in persuading us to accept solar energy, windmills and electric cars.

And there are a good many organisations and individuals who are now making very good money out of the climate change myth. They are helping themselves to vast quantities of pounds of public money.

Finally, the politicians who know that the oil is running out don't want to frighten us with that truth but are desperately keen to ensure that we all use fewer fossil fuels in order to preserve what remains.

4.

Electric cars are supposed to be the answer to all our environmental prayers. This is absolute tosh. The manufacture of electric cars requires massive amounts of polluting energy. And running them inevitably requires vast quantities of electricity – most of which is produced inefficiently using oil or coal.

Electric cars (which the witless Greens are so fond of driving and advocating) obtain their electricity from power stations. And where do the Greens and Little Greta think the electricity comes from? Well, gosh, golly and "shiver me timbers" most of the electricity for those nice electric cars is created by burning coal or diesel. Electric cars may not burn petrol or diesel directly but because the whole process of turning coal and diesel into electricity and then feeding the electricity into a car is a remarkably inefficient one, the sad truth is that electric cars require more fossil fuel per mile than traditional cars with internal combustion engines. So, anyone who wants to save the planet should buy a proper car and not a dirty, diesel and coal munching electric car. It is a plain, old-fashioned lie to claim that electric cars are 'zero emission vehicles'.

(Incidentally, I wonder how many fans of Tesla, the electric car company, know that the company deliberately stunts its batteries – reducing the range that drivers can travel between charges. If you own a Tesla motor car and want it to travel further, you have to pay the company an extra fee to have your battery upgraded. Isn't that nice of them?)

CCMs find it difficult to accept that many electric cars have higher lifecycle greenhouse gas emissions than petrol driven vehicles, and they forget that making electric cars requires massive amounts of energy. They forget that the factories required to build electric cars require a huge expenditure of energy. In Germany, Tesla has faced enormous protests from environmentalists over their plans to chop down over 200 acres of forest to build a new factory to make cars and batteries. This seems strange behaviour for a company apparently dedicated to saving the planet. Sadly, it seems that it is not at all unusual for those purporting to be concerned about the environment to be curiously careless in the way they treat the planet.

Fans of electric cars also forget that the huge quantities of rare earth metals such as lithium, rhodium and cobalt required for the batteries have to be dug out of the ground – using machinery which is powered with diesel or petrol. An electric car can require 10 kg of cobalt and 60 kg of lithium. And huge amounts of copper are needed too.

Recycling the parts of an electric car (and its batteries) requires more energy.

The fact is that electric cars have been around for 100 years and they don't have a future (unless we are all forced to buy them) because they are a truly terrible idea. Despite much publicity (and many subsidies) the only people who buy them at the moment are CCMs, people who want to look as though they care, and rich people who use them as second cars for trips within London where the drivers of petrol and diesel cars have to pay a surcharge.

5.

The climate change believers want to stop us using coal. And they want us to build more windmills. Companies digging out coal are being persecuted by just about everyone in search of brownie points from the CCMs (CCMs). Coal companies are even being denied insurance by big insurance companies pretending to be ethical. It is difficult to imagine anything more hysterical and hypocritical than insurance companies attempting to take the high moral ground.

Unfortunately, there is a problem.

You need to use a good deal of steel to build a windmill.

And you cannot make steel without coal. So, without coal mines there won't be any windmills.

Someone really ought to tell the CCMs about this.

6.

CCMs warn that if their climate change theories turn out to be correct, large areas of coastline will end up under water. What the CCMs don't realise is that the people living in those areas will be able to adapt their environment very successfully. Worldwide, 110 million people already live below sea level and have done so quite happily for many years. In London, for example, almost a million people live below the high water mark. Much of Holland is technically below sea level. Most of inhabited Vietnam is under water.

7.

The Environmental Agency in the UK claims that 'intense rainfall events' are more common today because of climate change. This simply isn't true. The highest rainfall in five minutes occurred in 1893. The highest rainfall in one hour occurred in 1901. The highest rainfall in 24 hours occurred in 1955. These figures all come from the Met Office. If the Environmental Agency cared about facts and science they could have easily obtained the figures before they joined the hysterical children who appear to be making government policy these days.

8.
We are constantly told that Iceland's glaciers are melting. But they aren't melting any more than they have been since the ninth century. The curiously named OK glacier, which pretty well disappeared in August 2019, had actually almost disappeared over 50 years ago. The fact is that some of Iceland's glaciers are now bigger than they were 1,000 years ago.

9.
It isn't difficult to fiddle the science if you want to convince naïve members of the public that the climate is changing. So, for example, if you have 100 temperature measuring stations spread around then you are more likely to see the lowest temperature for (say) 20 years than when there were just 10 temperature measuring stations.

10.
Idiotic bankers are forecasting a $12 trillion level of economic disruption as wind farms, solar panel manufacturers and others involved in the climate change industry take over from oil and gas companies.

This, of course, is absolute nonsense and simply goes to show how stupid bankers really are.

It is generally agreed by people who have been to school and who are not members of the British royal family, that the planet will need supplies of oil for decades to come.

There is no way that we will be able to manage without oil and gas for generations to come unless politicians intend to allow millions to die of starvation or cold.

Meanwhile, however, this massive unsubstantiated scare is a magnificent buying opening for those who want regular dividends and capital growth and who consider that big oil and mining companies are massively undervalued.

11.

A relatively easy solution to our energy problems would be to have lots of nuclear power stations. They're relatively safe and very efficient. Unfortunately, the Greens and CCMs don't like nuclear power and so we probably won't be having any of those. Oddly enough, though, the Greens and CCMs do seem to approve of chopping down trees (which we are all supposed to be planting), renaming the wood as 'biomass', shoving the 'biomass' into big incinerators and using the electricity produced.

12.

The climate change fanatics blame flooding on climate change and regard the fact that the nation has for decades been building approximately 10% of its new houses on flood plains as no more than an inconvenient coincidence. They also regard the fact that councils have failed to unblock drains or dredge rivers as irrelevant. Dredging is no longer fashionable and is inexcusably opposed by the European Union.

The falling numbers of salmon in the River Exe in the English county of Devon, for example, appears to have been caused by the deliberate blocking of several miles of ditches – thereby preventing fresh water from reaching the river – but CCMs are not interested in simple truths. Flooding, they believe, has absolutely nothing to do with the draining of peat bogs (nature's natural reservoirs) or the enthusiasm of farmers for chopping down every tree they can possibly find cluttering up their land. The CCMs prefer to blame the flooding on some mysterious aspect of climate change. 'It's the weather.' Even the chief executive of the Environment Agency has blamed 'the climate emergency'. Sadly, watercourses which are properly cleared and dredged do not flood but rivers which are controlled by the Environment Agency, and which are not cleared, do flood.

But the facts don't support these claims. Rainfall records have been kept in Lincolnshire since 1829 and these show that we are getting less extreme weather events than we used to get. The worst year ever was 1880. Other farmers who keep records have noticed the same thing.

The flooding problem has been exacerbated by the fact that local authorities have allowed thousands of houses to be built on flood plains. Today, one in ten new homes is being built on land that is known to flood. Moreover, tens of thousands of new houses are due to be built on flood plains in the next few years. When these homes flood, the CCMs will doubtless still blame global warming.

So many new houses are being built on flood plains that insurance companies are demanding that the Government provide financial support for the insurance of houses built after 2009 – proving that councils are still allowing builders to erect houses (and shops, offices and factories) on land that is known to flood.

13.

At the last British general election, the Labour Party promised to plant two billion trees by 2040. Presumably, no one in the Labour Party had bothered to work out that this would involve covering an area the size of Wales with nothing but trees. Or maybe the Labour party politicians realised that there was no chance of their winning the election and having to meet their promise.

14.

Celebrity climate change campaigners claim to be able to balance their private jet usage by planting trees but their claims are nonsensical. All that happens is that the celebrities trade their 'carbon credits' with forestry companies.

There are three problems.

First, the trees would have been planted anyway. Second, you need to do a lot of ploughing before you can plant trees – and ploughing requires tractors and fuel. Third, scientists from the University of Helsinki have shown that pines, birches and spruce actually increase the nitrous oxide in the atmosphere. Oh and the runoff from conifer forests is acidifying our rivers and making the planet less inhabitable.

15.

When I was young, car windscreens were covered in dead flies during the summer months. Motorists frequently fitted small plastic deflectors to their car bonnets in an attempt to prevent the flies smashing, bloodily, onto their windscreen glass. Today, this is no longer a problem. Moreover, there are very few butterflies around in fields and gardens these days, and fruit growers bemoan the dangerous lack of bees. The CCMs blame climate change but they're wrong, of course. The absence of bees, butterflies and other insects is entirely due to the popularity of insecticides and other chemicals on farms and in gardens. The amateur environmentalists are worrying about the wrong things.

16.

Britain's 2008 Climate Change Act (introduced and passed to please the EU) has impoverished and destroyed the British. The plethora of expensive and poisonous nonsenses resulting from the heavily subsidised climate change mythology (biofuels, electric cars, recycling absurdities, energy hungry windmills, pointless tidal energy experiments) have made energy so expensive that hundreds of thousands can no longer afford to keep warm. Tens of thousands now die every winter because our energy prices have been pushed up and up to satisfy the myths favoured by a bunch of unelected eurocrats who are supported by our craven and ever obedient politicians. In the UK, it is known that 4.5 million households struggle through the winter unable to heat their homes adequately. The actions of ignorant and hysterical environmental activists are making this situation worse. It is, perhaps, of significance that most of those who protest about climate change are mostly wealthy celebrities, comfortable, middle class 'warriors' or people on benefits who do not have to worry about energy prices.

17.

In 2020, a massive conference in Poland attracted 22,000 sanctimonious, self-indulgent hypocrites who flew in on yet another expenses-paid for booze up to discuss the usual myths and wishy-washy, pseudoscientific gibberish. Poland, we must remember, is a country which digs out a good deal of coal and has possibly the worst air pollution in Europe.

18.

Because of the EU's absurd and scientifically unsound obsession with climate change, local councils are not allowed to collect rubbish in the sensible, old-fashioned way. Councils are heavily fined if they do not force householders to put their rubbish into recycling containers. Residents in some areas are told that their rubbish must be out on the pavement before 7.00 a.m. and they will be fined if they put out their bins and boxes before midnight. No one in authority seems to care that the carefully sorted recycling rubbish is then sent to China to be dumped.

Councils have also abandoned their weekly collections and now collect rubbish once every two, three or four weeks. Not collecting rubbish regularly means that the rat population in Europe is soaring. And rats are getting healthier and bigger. Rats have up to 14 babies in a single litter, and the young rats are themselves reproducing within five weeks. The increase in the rat population is pretty well entirely dependent upon the amount of available food, and so it's a fair bet that the total rat population of Britain will be rising fairly dramatically in the coming years. Local councils don't seem to understand that rats are perfectly capable of eating their way into those disgusting plastic food containers which householders are encouraged to put outside their homes. Indeed, in towns and cities, the rats are now probably big enough to carry the containers away with them.

There are other problems in addition to the rats.

Councils are increasing their income by fining local residents who make mistakes. Last year, councils in Britain fined seven million people for failing to put their bins out at the right time, for not washing their yoghurt cartons properly or for mistakenly putting the right piece of recycling into the wrong container. What councils don't seem to understand is that the new system dramatically increases the spread of infection, with bin men transferring bugs from one recycling container to all the others they touch. If you have five or six containers then the risk is multiplied five or six times. The EU bin policies are spreading dangerous bugs. How many people wash their hands thoroughly when they have collected their variety of bins and plastic boxes from the pavement outside their home?

Unless home owners wear disposable gloves (made of plastic) when handling their bins, or wash their hands very thoroughly after doing so, they will have contaminated themselves and their homes with every possible lethal bug in the neighbourhood.

In an attempt to ensure that the 'go green' targets are met, several local councils in Britain are now insisting that wheelie bins are made out of green coloured plastic. So they are throwing out the old-fashioned, black rubbish bins and making green rubbish bins to save the planet.

19.

Has anyone thought what will happen when several million motorists arrive home in the evening and plug in their electric cars? Will the grid be able to cope? Will there be electricity left for heating and cooking?

In the past we have been told that the national grid has struggled to cope with the demand when fans have rushed to put on the kettle at halftime in big football matches.

And how, pray, will car owners who live in terraced houses or high rise blocks of flats manage to charge their electric cars? Who will pay for the installation of a couple of hundred charging points for every high rise tower? According to Scottish Power, the UK will need 25.3 million charging points if the Government's targets for getting rid of petrol and diesel cars are to be met. There are 30,000 charging points at the moment, so someone needs to build 25,270,000 more within the next 15 years at the latest. The cost will run into tens (or hundreds) of billions. And we will also need a good many new electricity generating plants building quickly too.

Maybe Little Greta has the answer to this thorny problem. No one else seems to have a clue about what to do.

20.
Because electric cars are silent there will doubtless be a good many accidents involving pedestrians. And that is before we allow our roads to be overrun with cars which are driving themselves.

21.
If we stop using fossil fuels then the Greens will have to give up using their beloved laptops and mobile phones and they will have to walk everywhere. All those lovely climate change conferences in wonderful locations will have to stop. And no electricity means that there won't be any Skype or video conferencing either.

It is one of the strange ironies that Greens are among the world's most enthusiastic users of social media and the most frequent travellers (and, therefore, users of fossil fuels). It apparently does not occur to them that until someone invents a solar powered aeroplane, they will be using up vast amounts of the world's depleting supply of natural energy.

22.
The Mekong, Yangtze and Ganges rivers are drying up. Salt water is surging up the estuaries. Naturally, the CCMs blame all this on man-made climate change. But they are wrong. Man-made dams have stopped the natural flow of water.

The CCMs never like the facts to stand in the way of their hysteria.

23.
Every year, the world uses 50 billion tons of sand and gravel. Huge building projects in China mean a massive demand for cement. China used more cement between 2011 and 2013 than America used in the whole of the 20th century.

Taking sand from beaches, coastlines and rivers changes those environments and results in reduced fish stocks and altered tide patterns. The sand which is used for building is the coarse sand which has been 'created' by water.

(The sand found in deserts, of which there is plenty, is too fine to be of any use for building projects and is of value only to the manufacturers of egg timers.)

24.

CCMs are inevitably pro-European Union. This is not a coincidence. These people are extremely gullible. They believe everything they are told by eurocrats and they are easily indoctrinated because they have little knowledge and less imagination. They support the BBC because they do not understand that the organisation is as bent as a nine euro note. They don't know how to think for themselves or question what they are told. They have been taught to regard democracy, independence, freedom, respect and dignity as outmoded concepts.

Children who have poorly developed imaginations because they have been brought up playing with computer games and who have been brainwashed by lobbyists and indoctrinated into believing in the EU, do not understand that their elders (and betters) loathe the organisation not because they are unduly patriotic but because they believe in freedom, respect and dignity.

The CCMs cannot bear to be contradicted and are quite unable to countenance that there could be any other point of view about anything. They control social media and therefore control public policies but eventually the mass of sensible people are rebelling.

Social media influencers are harmless enough when they are selling frocks and shirts but it seems to me that influencers such as Greta Thunberg, David Attenborough and Prince Charles are selling unscientific fear. It is not 'science' to say that climate change is happening simply because the BBC says it is. The gullible generation don't like discussing or debating.

25.

A government which robs Peter to pay Paul can always depend upon the support of Paul, said George Bernard Shaw. And, of course, Mr Shaw was right. Wealthy landowners tend to support the EU and to believe in the pseudoscience of climate change, because they receive massive farm subsidies which are handed out by the EU.

26.
The main reason to stop flying is nothing to do with climate change. Flying should be banned because of the way it spreads bugs around the world.

27.
Climate change protestors have tried to stop trains running by climbing onto their roofs. If they knew anything at all they would realise that the railways are the best way to move people about at low environmental cost. If the British Government really wanted to improve the environment and reduce reliance on oil, it would abandon its mad plan to spend £200 billion on the HS2 line designed to help northerners to cut a few minutes off train journeys to London and reopen all those railway lines (and stations) which were closed over half a century ago by the short-sighted Dr Beeching.

28.
The Met Office in Britain has forecast that the summer of 2070 will be five degrees warmer than normal. Since the Met Office isn't very good at forecasting what the weather is going to be like 24 hours ahead or, indeed, telling us what it is like at the moment, I am impressed that they have confidence in their ability to forecast the temperature 50 years ahead. Still, they will no doubt get extra brownie points from the Climate Change Fraudsters. Maybe they can work their way backwards and tell us, with a 50% chance of accuracy, what the weather is likely to be next weekend. Actually, it would be nice if they could tell us what the weather is likely to be like later on today.

29.
There is much talk that cliff falls are a new phenomenon caused by climate change. This is nonsense. On one part of the Devon coast, in 1790, heavy rain caused ten acres of land to slip 198 feet downwards. The land then moved 660 feet seawards to form an area known as Hooken Undercliffs. There have been no cliff falls of that size in recent decades.

    The one thing that would make a difference would be to ensure that all the bits of cliff which fall are left where they lie. Taking away the rubble, stone and sand merely accelerates the process of destruction. Leaving the fallen bits and pieces where they are helps to protect the remaining cliff face from the sea's perfectly natural attempts to claw its way inland.

30.
When Europe had a sunny couple of weeks, the climate change freaks abandoned their threat of global cooling and went back to calling it global warming. Sensible folk will prefer to call it simply 'unpredictable weather'.

    The truth is that there have been unusual weather events since the beginning of records. For example, back in 1709, Europe froze for several months. It was so cold that people skated on the canals in Venice. Church bells shattered when they were rung. It was possible to ride across the Baltic Sea on horseback. In England, the summers of 1911 and 1976 were far hotter, and for longer, than the summer of 2018 – currently being held up as an example of global warming. *The Guardian* newspaper claimed that the 'record breaking heat' of 2018 had forced people to wake up to the fact that 'something abnormal' was happening to the global climate. What typical *Guardian* nonsense. The fact is that 2018 was not the hottest year on record. It was a reasonable summer. That's all.

    Today, every unusual weather event is blamed on climate change. The utterly absurd extrapolations, exhortations, demands and instructions from the Intergovernmental Panel on Climate Change (the IPCC) would be laughable if they were not taken so seriously by so many otherwise sane seeming individuals. The IPCC has been making hysterical predictions for years, and sensible folk now take as much notice of their 'last chance to save the planet' scare stories as they did of the placards which used to warn us that 'The End is Nigh'.

Every time there is a weather related problem, it is blamed on climate change. If the weather is warm it is a result of climate change. If the weather is wet it is a result of climate change. If a blocked river floods it is a result of climate change. If the weather is a tad overcast or the air a little stuffy then we must blame climate change.

When we look back in history, we can see that storms and hurricanes and tempests and tsunamis have always been a part of our climatic history. Things appear worse today because the authorities insist on categorising every minor event as a major incident. Every breeze that rustles the trees is given a name. And afterwards no one remembers that Storm Greta hardly had the strength to move a leaf off a tree or that Hurricane Herbert turned out to be more of a Stiff Breeze than the devastation as advertised. The meteorologists will doubtless be terrifying us all with loads of false scares in the years to come – partly to justify their existence and partly to fit in with the global warming myth which is so popular these days. Every breeze, every snowfall, every warm day is now officially a sign of climate change. Everything must be exaggerated. A hot day becomes a heat wave. A windy day is a hurricane. Every storm is the Storm of the Century. Every warm summer is the Hottest Summer Ever. Virtually none of these claims is true. (In February 2020, Storm Ciara was widely described as the 'storm of the century'. It wasn't even a contender.)

But in truth our weather hasn't changed; it is our way of reporting it that has changed.

There has been snow in Scotland and the Peak District for centuries.

31.

Spurious climate scares have been common for generations. So, for example, back on 24th June 1974, *Time* magazine (that's the silly rag which made Little Greta their person of the year in 2019) announced the coming of another ice age. We are still waiting for that one.

Scare stories always attract readers, viewers and listeners and the gullible will always believe what they read, see and hear. I fear that the current, fashionable climate change madness was inspired by ignorant lunatics with access to social media and too much time on their hands.

Predictions and forecasts made by climate change 'scientists' have been woefully inaccurate – consistently.

Back in 2007, the WWF told us that we had five years to save the world. The Climate Change Hysterics told us that the English county of Cornwall would be a desert by 2010. In 2011, the International Energy Agency said we had five years to avoid Armageddon. In 2017, the United Nations said we had three years left and in that same year the International Energy Agency also said we had three years left.

Some of these merry doomsters are relatively cautious and merely claim that our planet will be unliveable within a generation. Others are far more specific. Greta Thunberg recently announced that we had eight years left to save the planet. I don't think she explained why it was eight and not seven or nine years before the Four Horsemen would ride into view in their electric cars. It seems to me bizarre that a relatively uneducated girl with no scientific background feels able to be so dogmatic. Is it at all possible that someone is feeding her opinions, I wonder? An American politician called Alexandria Ocasio-Cortez is more optimistic. Last year she said we have 12 years left before something will happen. In 2013, a Cambridge professor called Peter Wadhams said that we had until 2015 before all the Arctic ice disappeared. Mind you, he was optimistic compared to Gordon Brown who, in 2009, taking a tea-break from buggering up the British economy, told us that we had just 50 days to save the planet. And in 2004, the readers of *The Observer* were warned that by 2020, Britons would be living in a Siberian climate, though I'm not sure how they fitted that into the 'global warming' theory. Eleven years ago, Prince Charles said that we had eight years left to save the planet, so you might imagine that the heir to the throne would be hiding in a cupboard feeling rather embarrassed since there is clearly now no point whatsoever in doing anything to oppose the terror which awaits. However, Charles is made of sterner stuff than most of us and he is continuing with his scaremongering without allowing his past predictions to interfere with his latest proselytizing.

All this wild, scary stuff merely proves that the whole global warming/climate change thing is a hoax, an international scam of Brobdingnagian proportions. Despite the evidence the Mythmakers will doubtless keep going with their predictions. And, of course, making a prediction about the end of the world is a great way to get publicity and pick up more Twitter, Instagram and Facebook followers. The trick, it seems, is to pick a date a few years ahead and then hope that by the time we get there everyone will have forgotten what you said.

The CCMs remind me of that chap who, for 25 years, used to wander up and down Oxford Street carrying a board warning people to eat less protein.

It occurs to me that if Greta confined herself to walking up and down Oxford Street with a placard predicting the end of the world she would do far less harm.

32.

Politicians and journalists are as potty over electric cars as they were over biofuels. The British Government now insists that car manufacturers must stop making diesel or petrol driven cars by 2035 at the latest. If we want to be mobile we will have to buy electric cars. Biofuels will cause millions to die of starvation – short of food because land has been used to grow crops for fuel. And by using up all the available electricity, electric cars will cause millions to die of the cold.

33.

Why have none of the enthusiasts who promote the concept of climate change so vigorously and profitably not mentioned the fact that our climate will change as Britain becomes ever more crowded – thanks to absurd immigration policies foisted on us by the EU?

It is well-known that heavily populated areas of the country are warmer than relatively unpopulated areas, and so it seems to me pretty obvious that as the nation's population rapidly increases so the temperature will also increase (and the levels of air pollution will also rise).

If the climate change 'experts' really believe what they say they should be campaigning vigorously against immigration.

34.

It is a delight to me that virtually no members of the general public are taken in by the CCM's nonsense. Sensible people no longer trust anything 'they' say because 'they' lie about everything – partly to keep us all terrified, and therefore compliant, and partly for specific political or commercial purposes.

Notably, there is no reliable scientific evidence to support the persistently hysterical judgements proffered by important and allegedly independent scientific bodies such as the Bank of England and the BBC.

(It's interesting to note, incidentally, that the Bank of England now seems to be staffed exclusively with climatologists. The Bank of England has a lot to say about our weather but seem utterly useless at forecasting economic trends or saying anything sensible about the economy. And why, if the Bank is so concerned about wasting energy, are billions of perfectly serviceable £20 notes to be burnt and replaced with new plastic ones?)

The most hysterical CCMs allege that our planet will be unliveable within a generation. There is not one jot of evidence for this absurd claim.

35.

In February 2020, Boris Johnson promised to phase out coal power by the year 2024. He did not say what would replace the missing coal power. Presumably, Britain will simply chop down more trees and burn them instead.

36.

The alternatives to fossil fuels are frighteningly inefficient. For example, it takes more energy to make a windmill than the windmill will ever produce. So, the more windmills we produce the more energy we waste. Greens are madly enthusiastic about wind power – and they pressurise governments to give subsidies to wind power installations. (These subsidies mean that the poor and the elderly pay most of the cost of wind power.) But windmills are not just ugly and dangerous to birds – they are also horrendously inefficient. They require vast amounts of energy to build. And they are never energy efficient. Indeed, they are negative energy sources. Wind farms and solar farms are examples of entirely pointless rural vandalism; tributes to the self-serving sanctimoniousness and rank hypocrisy of an ignorant generation.

37.

Electricity, though very nice and useful stuff, only provides about 20% of our energy needs. The other 80% comes from nasty old gas, oil, coal and nuclear power. And it is nigh on impossible to increase that proportion. An awful lot of people rely on gas for their central heating and cooking. If all those people are forced to use electricity for heating and cooking then there is going to be a great shortage of electricity because we are already using up every drop of the stuff that we can make.

38.

Renewables such as solar energy and wind power provide only a tiny portion of our current electricity needs. We would need to carpet the countryside with solar farms, wind farms and huge log burning incinerators to increase that proportion significantly. And without the subsidies which are currently paid by consumers to rich landowners, the electricity produced would be horrifically expensive.

It's important to remember that because of the subsidies which have been paid to rich farmers operating wind farms and solar farms, ordinary people have had to pay more for their heating and many thousands have died in the cold weather.

Is that supposed to be a good thing?

The organisation Ofgem, which regulates gas and electricity markets in the UK, has confirmed that the European Union's various 'energy' schemes have already added 6% to the average energy bill in Britain.

39.

CCMs and Greens always forget to mention that when there is no wind, the windmills have to be turned (using electricity) in order to stop them seizing up. Solar energy panels don't require energy to keep them working but they require so much energy to build, transport and install that they too are negative producers of energy. (In other words, they require more energy to make and maintain than they will ever produce).

40.

There is no evidence that it is a bad thing if the earth is getting hotter.

Rising tides would be bad for people with beachfront properties in some parts of the world.

But if there is more carbon dioxide in the atmosphere, there will be more plants on the planet and then there will be less starvation. Surely we can all agree that would be a good thing?

41.

The Government is introducing legislation to prevent home owners burning logs in their hearth. Log burners will be banned. And yet by far the biggest 'renewable' source of energy comes from burning biomass. And biomass, as we know, is wood from trees. The log burner manufacturers would be wise to rename their log burners – calling them 'biomass burners'. Then they would probably get a grant. The main losers from this legislation will be country dwellers who rely on open fires for heating, light and cooking when overhead electricity supplies are disrupted. Since the Government is also planning to ban gas appliances, this will be a serious blow to millions.

42.

Electricity provides us with a fifth of our energy requirements. Renewable sources of energy such as wind power, solar power and burning trees (which is by far the biggest 'renewable' source of energy) can only produce electricity, so without fossil fuels we will have to survive on sources of energy which provides just 20% of our current needs.

We get the other 80% of our energy from oil and gas and, sadly, climate change hysterics have not yet realised that you can't make oil or gas from windmills or solar panels or by burning trees.

However, since renewables only provide a quarter of our electricity (and it will be difficult to increase that figure without increasing the amount of wood we call biomass and then burn) we will have to survive on just 5% of the energy we use at the moment (that's a quarter of the 20% of our energy we obtain from electricity).

This is a bit of a problem because we can hardly cope on the energy supply we have now.

If we give up using oil, gas and coal we will have to give up all forms of powered transport (including cars, planes and ships); all forms of entertainment which require electricity (e.g. television, radio, computers, mobile phones and so on); all forms of heating; all factories which make things; all mechanised farming and all fertilisers; all hospitals, medical treatments and all drug production. And the real bummer is that the small amount of energy we have will be needed to maintain and service our solar panels and our windmills. In the unlikely event of there being any energy left over, we may be able to boil a kettle and make a cup of hot water, though there won't be any tea leaves, milk or sugar to put in the hot water. The good news is that our inability to use tractors and fertilisers will mean that most of us will starve to death, so we won't mind too much.

43.

If the Greens have their way, our planet will plunge into the biggest war of all time. The survivors will be those countries which retain fossil fuels and use them to manufacture armaments and to make and fuel bombers and tanks. The citizens of countries which decide unilaterally to rely on renewables will die.

Our energy policies will define our future.

44.

Oil companies have pretty well stopped exploring for more oil, partly because of very vocal opposition from idiots who seem to think cars, lorries and planes can all run on solar power but mainly because climate change campaigners have pressured the banks into refusing to lend money to companies exploring for oil. In February 2020, for example, a bunch of shareholders calling themselves The Investor Forum (which apparently represents some of the UK's biggest investors) complained that Barclays was the biggest financier of fossil fuel companies in Europe. The investors have apparently demanded that Barclays adopt a stricter climate change policy – and stop lending money to energy companies such as those trying to find oil. The investors would presumably prefer us to fit motor cars, lorries and aeroplanes and ships with log burning stoves so that they can move about using 'biomass'.

Without new sources of oil, the existing oil supplies will run out sooner and that will be that.

45.

Renewable energy sources will not be able to replace oil and coal in the foreseeable future. The most reliable estimate currently available (from the International Energy Agency which 'works with countries around the world to shape energy policies for a secure and sustainable future') suggests that by the year 2040, the world will still be obtaining a little over 5% of its supply of energy from hydroelectric sources, wind-power, solar power and burning the trees we are supposed to be planting. Most of the other 95% of our energy will have to come from oil, gas and coal. A small proportion will come from nuclear power. That, remember, is the forecast from the International Energy Agency. It's a fair guess that they know more about our energy needs than Prince Charles or young Swedish girls.

The EU and the British Government are committed to getting rid of coal, oil and gas, so it's no exaggeration to say that in just over two decades, Britons will have to make do with 5% of the energy we use at the moment.

46.

The EU has stated that a fifth of all energy must be 'green' (obtained from biofuels) by 2020. In America, a proposed energy package will require 15% of all transport fuels to be made from biofuel by 2022, proving that the Americans are also stupid but not quite as stupid as the eurocrats. The targets are as absurd and as harmful as they are arbitrary.

We must, say the eurocrats, have green fuel, green cars, green rubbish collections, green homes, green factories and green aeroplanes. The EU even says that we must soon generate 20% of our electricity from renewable sources. It is this nonsensical policy which will result in our countryside being covered with wind turbines. Since windmills are woefully inefficient, this will add to our looming energy crisis.

Naturally, in order to ensure that their goal is reached, the EU is subsidising the growing of biofuels. This makes global warming worse, results in a lot of people starving to death and makes big money for a few people. Companies growing biofuels are making huge profits.

The European Commission, father and mother of the straight banana and the perfect cucumber, and a safe but wildly overpaid working environment for many of the world's most blindingly stupid people, called for biofuels to replace 10% of all petrol by the year 2020 and it ruled that to reach its target, 25% of all European arable land should be turned over to ethanol production.

The consequences of this madness are seemingly unending.

Farmers who used to grow soybeans are now growing corn because they can either sell it as food or sell it to the oil companies to mix in with their petrol. Rapeseed crops are now seen everywhere in Europe as farmers abandon growing barley and wheat. (The disappearance of barley is, of course, why beer prices are rising. And will continue to rise.)

I wonder how many farmers know (or care) that rapeseed oil produces lots of nitrous oxide gas – which creates even more global warming than carbon dioxide? I wonder how many eurocrats know that you can't send ethanol along the usual gasoline pipelines because it is corrosive and picks up impurities. (Don't ask what it does to petrol tanks.) But I'm prepared to hazard a guess that not many know that this means the darned stuff has to be transported in tanks by rail. And that means building more suitable containers and more trains. The energy cost of all this construction is phenomenal. To accurately assess the energy value of biofuels we need to consider a whole raft of extra costs – including the cost of growing the extra food to feed the farm workers growing the biofuel. The labour and energy costs for extracting oil are very small.

Those who believe that ethanol offers answers to all our problems are magnificently, technicolour stupid. Those who believe that ethanol offers any answers to some of our problems are just plain stupid.

The lunatic belief in ethanol has increased the price of food around the world and is boosting the price of farmland. The makers of Italian pasta, Mexican tortillas, American corned beef and German beer have all warned that the price of their products will have to rocket. Food conglomerates have issued a series of profit warnings. The ethanol promoters don't seem to care that their bright new idea will result in a massive increase in worldwide starvation. Forty per cent of the population in Swaziland were starving recently while their Government was allowing agricultural land to be used to grow enormously profitable biofuels.

The pro-ethanol strategy (as they like to think of it) is eye wateringly wasteful. It's so stupid that it's difficult to say how stupid it is. But I will try.

The EU's biofuel policy is like trying to replace the world's shrinking glaciers by making ice cubes in a series of refrigerators, carting the ice cubes by lorry to the site of the shrunken glaciers, and then hoping that the ice cubes will repair the damage.

Here are several good reasons why biofuels won't work.

First, creating ethanol from corn will use up vast amounts of our corn supply. The corn needed for one tankful of ethanol could feed one person for a year. In Iowa in the USA, 25 ethanol plants are now operating, and 30 being built or planned. Once built, these plants will consume half the state's crop of corn. The prices for sugar, corn and wheat are rising because we are using these foodstuffs for fuel. A huge chunk of world grain consumption is now going into American petrol tanks. (At a time, it should be noted, when America has already become a net importer of food.) This will result in greater starvation in a world where there is a continuing and developing food shortage. The global lack of fresh water (caused by the increase in the number of bathrooms being built, the increase in the number of people on the planet, the increasing demands of providing water for increasing numbers of cattle, the increasing pollution of our water supplies and the use of vast amounts of fresh water to wash out jam jars, baked bean tins and yoghurt cartons) will make growing crops even more difficult. And will also result in a greater shortage of water and higher prices for basic foodstuffs.

Using rape-seed and sugar beet instead of corn doesn't help because (and this will doubtless come as something of a surprise to politicians and bureaucrats) you still need land to grow the rape-seed and sugar beet.

And when you've grown the stuff, you still need to use lots of energy to harvest it, transport it (often over huge distances) and turn it into ethanol. The dullards at the European Commission probably imagine that you plant some sugar beet seeds and wait for a nice neat row of petrol pumps to pop up a few weeks later.

Second, it costs more in terms of energy to get the fuel than it delivers. There are energy costs in building and running tractors, and moving the corn around. You have to include the energy needed to plant the corn, water it, harvest it and turn into alcohol. Oil is needed to make the fertilisers and pesticides the farmers use. Numerous experts have shown that producing ethanol actually costs energy. According to David Pimentel, a professor of ecology and agricultural science at Cornell University in the USA, 129,600 British thermal units of energy are used to produce one gallon of ethanol. And one gallon of ethanol provides just 76,000 BTU of energy. Another independent expert has worked out that around 131,000 BTUs are needed to make a gallon of ethanol but that gallon will only produce 77,000 BTU. These figures mean a loss of over 50,000 BTU for every gallon of ethanol produced.

The most important thing to remember about fuels is the energy output to energy input ratio. This measures the amount of work you have to do in order to obtain the fuel. The energy output to input ratio for oil is magnificent. It varies between 30 to 1 and 200 to 1 depending on where the oil is. If it's easy to get at then the ratio is probably closer to 200 to 1. Ethanol is very different. It depends on plants which have to be grown afresh every year. To grow plants to turn into ethanol you have to plant seeds, harvest the crop, take the crop to the refinery and then start work on it. You need a tractor to plant the seed, a tractor to harvest the crop and a lorry to take the crop to the refinery. It takes fuel to create the fuel. Oh, and then there's the fertiliser, the herbicide and the pesticide the farmer uses. Where do those come from? Oil. The bottom line is that to obtain ethanol we have to use up energy to make energy. We get less out that we put in.

Third, almost all biofuels cause more greenhouse gas emissions than conventional fuels if the pollution caused by producing these so-called 'green' fuels is taken into account (which it obviously has to be). In February 2008, two studies published in the journal *Science* took a broad look at the emissions of the effect of converting huge amounts of land to grow biofuels. Scientists concluded that destroying natural ecosystems – whether they are rain forests in the tropics or grasslands in South America – increases the release of greenhouse gases into the atmosphere because the ecosystems are the planet's natural way of dealing with carbon emissions. Plant-based fuels were originally claimed to be better than fossil fuels because the carbon released when they are burned is balanced by the carbon absorbed when the plants grow. But even that was simplistic and wrong. The process of turning plants into fuel causes its own emissions because biofuels have to be refined and transported before they can be used.

The reports in Science journal showed that the clearance of grassland releases 93 times the amount of greenhouse gas that would be saved by the fuel made annually on that land. The scientists concluded that it doesn't really matter if you clear rain forest or scrub land – you are making things worse. In its usual pitiful way the EU has attempted to deal with this problem by proposing regulations stipulating that imported biofuels cannot come from land that was previously rain forest.

Brilliant.

What the eurocrats obviously didn't work out is that when the EU buys biofuels which have been grown on ordinary farmland, the price of food will rocket and so rain forests will be cleared to grow food so that people in Third World countries can have a little something to eat occasionally.

The stupidity of eurocrats sometimes leaves me open mouthed with amazement.

Indeed, the EU is already way behind events.

Farmers in Brazil have already been busy tearing down rain forests so that they can grow more soybeans. Naturally, the CCMs are getting cross about the Brazilians wanting to eat.

Fourth, pollution from ethanol could create worse health hazards than gasoline – especially for sufferers from respiratory diseases such as asthma. Ethanol burning cars increase the level of toxic ozone gas in the environment. This will mean that the atmosphere will become much more dangerous. Pollution from ethanol is more dangerous than petrol because when it breaks down in the atmosphere it produces considerably far more ozone than petrol does. Ozone is corrosive and damages the lungs. (Ozone is so corrosive it can crack rubber and destroy stone statues). Other substances released from use of ethanol as fuel include benzene, formaldehyde and acetaldehyde. All these are carcinogens.

Fifth, the search for so-called green fuels, with which the CCMs claim we can all prevent climate change and save the environment, is resulting in some strangely destructive behaviour. So, for example, the rising demand for palm oil (an ingredient in biodiesel) has led to tropical forests being cleared in vast areas of South East Asia. People are now chopping down hardwood forests in Asia in order to grow palm oil to take to the USA for the biodiesel industry. This is being done in order to keep petrol prices down so that Americans don't have to change their way of life.

And, in turn, means that there are fewer trees to get rid of all the carbon dioxide being produced by the tractors and lorries used by the farmers planting the corn, the sugar beet and the rape-seed.

Clearing rainforests to increase the land available for the cultivation of palm oil is bringing ecological disaster for countries such as Indonesia and Malaysia. Indonesia, a key palm oil producer, now has the worst carbon emissions level per head of population thanks to the fact that forests have been cut down to make room for palm oil production.

The United Nations has predicted that the natural rain forest of Indonesia will have disappeared in just 15 years' time, because of the planting of palm oil to turn into biodiesel so that Europeans (and Americans) can keep on driving their motor cars and taking long flights. Plantations of oil palms for biodiesel have been held responsible for 87% of deforestation in Malaysia. The fuel planters encouraged by the EU are ploughing up the planet. As I write, their bulldozers are busy in Africa and South America as well as Asia. The environmental damage seems endless. Sugar cane being grown in huge amounts degrades soils and causes pollution when fields are burnt to get rid of stubble and destroys wildlife.

The United Nations has also reported that biofuels use up vast amounts of water (a commodity that is becoming increasingly scarce). And mono-cropping (growing the same crop year after year) increases pests and has a negative impact on soil quality.

The bottom line is that bioethanol and biodiesel are a huge scam which will increase not reduce greenhouse gas emissions. Indeed, biofuels may well be the final nail in the planet's coffin.

It's about time the self-righteous nonsense preached about ethanol was put to rest. There is nothing remotely green about biofuels such as ethanol. It is a grotesque myth that biofuels are carbon neutral and will help with the oil shortage.

Biofuels are promoted as planet friendly. But ethanol (from corn or sugar cane) and biodiesel (made from soybean or palm oil) are no answer to any of our problems. The only people benefiting are the motorists who are able to keep driving their cars and feeling good about themselves, the governments who are able to keep collecting the taxes raised on those fuels and the corporations growing and selling the crops which are being used to make biofuels.

The only good thing to come out of all this is that OPEC has threatened to cut oil production if the West continues to use more ethanol. This will increase oil prices but it may help ensure that the oil lasts a year or two longer.

The American Government and the EU are spending billions of taxpayers' dollars and euros subsidising the biofuel industry. And yet the OECD has calculated that it would take 70% of all Europe's farmland to supply enough biofuels to save 10% of the oil currently used in transport.

Will anyone take any notice of all these truths? Will the Americans stop their mad race to replace oil with biofuel? Will CCMs change their minds?

Of course not.

For one thing the American and European farming lobbies are far too powerful to be ignored. And they are enjoying massive profits from the biofuel boom.

And for another, the Americans are desperate to weaken the economies of oil-rich countries such as Iran, Russia and Venezuela.

All around the world experts believe that the EU Commission's absurd proposal that biofuels should account for 10% of road transport fuels is dangerous nonsense. Even the Commission's own in-house science institute has questioned the policy and a London based economic consultancy, Europe Economics, has argued that biofuel targets amount to a 'form of state support for an environmentally and economically harmful activity designed to consolidate existing price support mechanisms for vested interest groups, mostly farmers'.

The economists say the target would lead to annual subsidies to the biofuels industry of £8.2 billion and possibly as much as £17 billion by 2020. This compares with the £30 billion spent each year on the EU's Common Agricultural Policy.

'It is clear that a European biofuel industry cannot be viable without political support by means of tariffs and a very high level of subsidy,' say the economists, whose study was commissioned by Open Europe, a UK think-tank.

The bottom line is that the EU policy on biofuels will drive food prices up even faster than they are already rising. It is estimated that the EU's biofuel policy will add £750 a year to the average family food bill.

Using food to make fuel has exacerbated the world's starvation problem and the Greens have condemned millions to death by starvation by campaigning for yet more food to be turned into fuel for rich people.

Celebrities who claim that they are offsetting their private jet usage often do so by helping to pay for crops to be turned into biofuels. These celebrities are managing an evil double whammy: their private jets are using up precious oil and damaging the planet and by turning food into biofuels they are causing thousands to starve to death.

47.

Electric car owners have been proved to be more selfish on the roads than drivers of proper vehicles. Electric cars have been proven to be worse for the environment than diesel or petrol vehicles but drivers still exude an air of smug superiority.

48.

The CCMs have at last cottoned on to the fact that if people stopped eating meat (or at least cut down their consumption) then the environment would be healthier. I wonder why it took them so long. It has been clear for years that all those belching and farting cows do considerable damage to the atmosphere. And they require vast amounts of water.

And, of course, we also know that if people stopped eating meat there would be plenty of land to grow food for the millions who are starving to death. So a widespread vegetarian lifestyle would end starvation.

Finally, it has been proven beyond doubt that eating meat causes cancer and is, indeed, as big a cause of cancer as tobacco.

But the Government won't officially recommend a meat free diet because (it says) it doesn't want to interfere in our lives and our freedom to eat what we want.

That's odd.

Because the Government has announced that it is going to tell food companies to make smaller pies and smaller pizzas. Is that not interfering?

The truth is that the Government is frightened to tell us to cut our meat consumption because it is frightened of farmers and the meat industry – a deadly bunch who are more powerful than the tobacco industry ever was.

49.

I have never been impressed by the intelligence of global warming conspiracy theorists. In 2019, a bunch of climate change activists spent much of the day painting silly slogans on public buildings in London. Removing the paint required much wasteful expenditure of energy. Having defaced a chunk of central London, the protestors then glued themselves to the gates at the entrance to Downing Street. And the police then compounded the stupidity of the whole pointless exercise by removing them. They should have left them glued to the gates for a couple of months.

Thinking, sentient members of the public are fed up with the childish, anti-social behaviour of ill-informed protestors. The activities of hysterical lunatics in London in 2019 cost the police more than £40 million (which could have been usefully spent fighting crime and other forms of terrorism), wasted much energy and caused serious disruption to law-abiding citizens going about their lawful business.

50.

Public figures all like to portray themselves as 'green' and 'environmentally aware'. When asked about the fact that he has three Hummer vehicles, the Governor of California, Arnold Schwarzenegger, announced that two of the vehicles had been converted to take biofuel. Managing to say this with a straight face must have required more acting skills than I realised he had.

51.

CCMs blamed a Californian forest fire on global warming. I do wish these idiots would do a little research before opening up their laptops and firing off their screams of anguish. The average annual acreage of American forest burned is now around 6.6 million. Back in 1928, the average annual acreage of American forest lost to fires was 41.7 million. I realise that millennial CCMs aren't terribly good at maths but I am pretty confident that 41.7 is a bigger figure than 6.6.

Similarly, forest fires in Australia are nothing to do with climate change.

52.

Figures recently released by Public Health England show that for the last 20 years the estimated number of deaths from hot weather in the

UK has run between one for every 100,000 citizens and six for every 100,000 citizens. But the estimated number of deaths from cold weather runs between 59 and 82 for every 100,000 people. Moreover, the official estimate is that if temperatures climb in line with the worst and most outlandish projections then the estimated deaths in the 2080s will be 7 to 21 per 100,000 from heat and from 28 to 56 per thousand from cold. The cold will still be a greater threat than the heat.

It is clear from these (official) figures that the biggest problem facing Britain is the high price of heating supplies which have already been pushed to unbearable levels by absurd and scientifically unsound climate change policies.

If we stop using fossil fuels, and stick with windmills, solar panels and incinerating trees in huge biomass burners, then the number of people dying from the cold will rocket. And most of those dying will be old.

Today's young campaigners don't seem to give a fig about the elderly. But I wonder when they will realise that today's 12-year-olds will be pensioners in 2080. And they will then be killed by their own stupid energy policies.

Green energy policies, inspired and controlled by the European Union and silly 12-year-olds, have done nothing to improve the planet but they have already pushed up the prices of energy to such a level that hundreds of thousands of pensioners have to choose between eating and keeping warm.

53.

Aviation only produces as much carbon dioxide as the world's computer data storage centres. All those banks of servers, upon which social media campaigners share their global warming nightmares, burn up vast amounts of electricity.

So if the CCMs really want to save the planet they should throw away their laptops and mobile phones and close down their social media accounts. They should stop tweeting, face-booking and instagramming!

Fat chance.

The CCMs love their self-important screeching far too much to set aside their iPhones and do something useful with their lives.

54.

The car industry is already struggling to find enough cobalt, lithium and to make batteries for electric cars. The planet's supply of these goodies is very limited. Oh, and there is, of course, a massive demand for copper for all the wiring. The mining required to dig up these elements requires lots of huge fuel guzzling equipment.

As an aside, half of all the cobalt needed is mined in the Democratic Republic of Congo. I suggest that CCMs do a little research into the Democratic Republic of Congo.

The price of the materials required for electric car batteries is going to soar and the mines mean much despoliation. The fans of electric cars never seem worried about this. Nor do they campaign to leave the cobalt and the lithium in the ground as they do with oil.

Many of the rare earths used in the manufacture of electric car batteries are dug out of the ground by children as young as seven. Sanctimonious electric car buyers will doubtless be delighted to know that they are providing work for so many under 12s.

55.

Hypocrisy among CCMs comes in many flavours.

There are the celebrities and royals who insist on flying to climate change demonstrations.

And there are the children who insist on being driven to school and then spend the day playing truant and supporting Little Greta's nonsensical campaign.

56.

Celebrities, politicians and other hypocrites who fly around the world but then tell the rest of us about the dangers of climate change respond to critics by claiming that they have purchased 'carbon offsets' to balance their carbon 'footprints'.

This is not a new idea.

In 2007, the British Government said that everyone, including politicians and civil servants, should have their own carbon swipe cards forcing them to take more care of the environment by offsetting their carbon emissions.

You will not be surprised to hear that the whole daft business of trading carbon emissions is an EU project. Offsetting carbon emissions is a sick joke; a perfect example of hypocrisy. What is the point? Surely if we are doing anything we should be reducing emissions?

Politicians and millionaire pop stars buy carbon emission offsets in order to 'balance' their flying around the world in private planes. (Private planes do far more damage, and use up far more valuable oil, on a per-person basis than the cheapest commercial airlines which cram in as many passengers as possible.)

But buying carbon offsets doesn't help the planet at all.

Carbon trading is practical hypocrisy.

No wonder the EU is an enthusiastic supporter.

The European Union, the world's largest and most corrupt fascist organisation, set up its own carbon trading programme called the Emissions Trading Scheme and as you might expect the European Union's scheme is of no benefit to the environment, the planet or you and me. As you might equally expect it is, on the other hand, of enormous benefit to the large companies which are responsible for polluting the planet.

Officially, the EU's carbon trading scheme was set up to encourage dirty power stations to switch to cleaner forms of energy but in practice the scheme has allowed the dirtiest polluters, the companies really responsible for global warming, to push up their bills and increase their profits without lowering the level of their emissions.

Even industry experts admit that the EU's carbon trading scheme has been a windfall which has allowed the big power companies to increase their profits massively. Way back in January 2008, it was announced that thanks to the EU's scheme, energy suppliers would be given a bonanza windfall of £9 billion.

You might have thought that if the EU really wanted to cut pollution, it would have introduced strict fines for companies which produce a lot of carbon dioxide. But the EU bureaucrats didn't want to do anything quite so logical, sensible, simple or effective.

The EU's scheme was designed to give permits to major electricity producers and manufacturers allowing them to produce a fixed amount of carbon dioxide every year.

The plan was that any company which reduced its allowed ration of pollution could sell its unused permit to pollute on the open market. A company which exceeds its allowed level of pollution had to buy an extra permit to pollute.

Anyone with functioning brain tissue would have forced companies to buy their permits to pollute. This would have given them a real financial incentive to reduce their level of pollution.

But the EU (supported and endorsed let me remind you by all three major political parties in Britain) succumbed to pressure from the polluters and handed out the permits free of charge.

In case you thought that might be a misprint I'll repeat it: the EU gave the power giants their permits (and the billion pound windfall payments) entirely free of charge.

What the witless idiots at the EU presumably hadn't realised was that the minute the big electricity producers had their permits, they would cut their output of electricity in order to reduce their level of pollutants. They would then be able to sell their spare polluting capacity to other companies. Having cut the amount of electricity they were producing, the big electricity companies were then able to put up their prices.

Double whammy!

A bunch of the world's biggest polluters made extra money by charging more for their electricity. And they made extra money by selling off part of their permit to pollute. (A permit which, remember, the EU had given them free of charge.)

According to the European Commission, the result of the EU's policies was that electricity prices needed to rise by 10% to 15% across Europe. And the capital commitment costs of meeting the EU's renewable energy targets were estimated at around £3,000 per household.

57.

As far as individuals are concerned, the idea of carbon offsets is that when you fly to, say, Bermuda to stay in a pop star's home, you invest in a project which either removes some carbon dioxide from the atmosphere or prevents some carbon dioxide being put into the atmosphere. There are now many small, and rapidly growing firms, offering a variety of such schemes.

Most of the people who 'offset' are air travellers – usually celebrities. This is probably because it's fairly easy to isolate the damage done by a flight and probably because it is a fairly inexpensive way to feel good about yourself.

Do these schemes work? Are they really going to make a difference?

The answer, I fear, is a fairly loud No.

Some projects (planting trees for example) would have taken place with or without anyone paying a carbon offset 'guilt' fee. As far as the planet is concerned nothing whatsoever is gained by giving the tree planter an extra fee for the right to his 'carbon offset'.

And there are persistent rumours that some unscrupulous individuals may sell the carbon value of their trees to numerous buyers. Since there is no register of who is paying for what it is perfectly possible that some slightly bent Arthur Daley character in some far off land could be selling and reselling the carbon offset value of the tree he planted many, many times.

No one knows.

Nor does anyone really know the damage done by a flight or the value of a tree to the planet.

When the magazine *Nature* asked four offset firms for the carbon dioxide emissions per person of a return London to Bangkok flight, they got four different answers – varying from 2.1 tons of carbon dioxide to 9.9 tons of carbon dioxide.

And, in practice, of course, the damage done depends on the number of people who were on the flight.

Fly by private jet (as celebrities and royals are likely to do) and you are obviously doing far more damage than if you are crammed sardine-like into a charter flight.

The other problem is that it is pretty well impossible to carbon offset all the terrible things we are doing.

To offset the UK's annual emissions total of carbon dioxide, we would have to plant and maintain forever a forest the size of Dorset. Every year. And whenever a tree was cut down another one would have to be planted.

(Remember that trees are cut down to produce 'biomass' which is then burnt to provide electricity. Biomass is the largest renewable source of electricity – far exceeding the amount of electricity produced by windmills or solar panels. CCMs are very keen on biomass because no one has yet told them that biomass is the same thing as wood and that wood comes from trees.)

The bottom line is that buying carbon offsets is just an easy way for celebrities, politicians and royals to feel good about themselves without having to amend their travel plans or slum it by travelling on a crowded commercial jet as opposed to a private jet.

And, it's a great way for people to make money.

It's all a bit like the system of 'indulgences' which was in vogue during medieval times. Sinners haggled with and then paid a corrupt priest a fee to absolve their sins.

The EU's carbon emissions trading scheme (ETS) has been severely criticised by economists who have pointed out that the most cost-effective way of reducing carbon emissions is to introduce an economy-wide carbon price and that the best way to do this would be to abandon the carbon scheme and to tax the consumption of primary fuels in proportion to their carbon content. That would, of course, be far too simple for the EU to consider.

The EU admitted that their crackpot policies will cost every household in Europe an average of £520 a year. 'But that's OK', said a spokesman. 'It's just the cost of three tankfuls of petrol a year.'

I don't know about you but I've yet to see a motor car that takes £173 worth of petrol. But I expect eurocrats are used to driving round in very large, very expensive limousines.

Incidentally, when Ryanair introduced a voluntary carbon offset programme, inviting customers who flew to make donations, just 2% of their customers made a contribution. The other 98% either didn't think much of carbon offset schemes or weren't taken in by the global warming myth.

58.

One tree can absorb as much carbon in a year as a car produces during 26,000 miles.

So, if you are a driver and you cover 10,000 miles a year (an average sort of motoring) then you will be in balance if you plant a tree every couple of years.

59.

Journalists make two mistakes about electric cars.

First, they seem to assume that the electricity they require is obtained by magic. It isn't, of course. Most of the electricity used to run cars is made from coal or oil.

Second, they promote electric cars as being cheaper to run. This isn't true. I saw an article today in which the writer pointed out that an electric car costs only £35 to charge at a typical public charging point – 'less than half the cost of a typical tankful of petrol or diesel'. What the writer forgot to mention is that an average electric car will go around 70 miles (with a tailwind) whereas an average petrol or diesel car will travel at least five times as far. It has been shown that if you buy a Nissan Leaf (which is an electric car) and keep it for ten years it will cost you a total of £31,435 if you drive it for 10,000 miles a year. But if you buy a Nissan Micra (a petrol driven version of a similar car) and use it the same amount, it will cost you £28,630. It must also be remembered that the electric car is heavily subsidised and includes a government grant of £3,500 and whereas the owner of the petrol driven car has to pay a total of £1,450 in road tax, the driver of the electric car pays nothing towards road maintenance and building. So the electric car turns out to be £6,755 more expensive than the petrol driven car.

Electrical cars are merely expensive, practical hypocrisy.

60.

It is an undeniable truth that electric cars are inefficient and more polluting than traditional vehicles (if you count their manufacture – which you must) and that their use will spark a massive shortage of electricity for cooking and heating.

People who use electric cars are selfish cheapskates who are damaging the environment – and their fellow citizens – so that they can appear self-righteous.

But there's another problem no one talks about.

In a few years' time we will have to deal with all the dead batteries taken from the electric cars on our roads. There will be millions and millions of batteries on a huge electric car battery mountain. And that's just the batteries from motor cars. Imagine the size and number of the batteries required to power lorries, aeroplanes and ships.

The cars sold in 2017 will produce 250,000 metric tons of battery pack waste. If all goes the way the Government wants it to go, there will be 5,000,000 tons of battery pack waste to dispose of in 2027. And more in 2028 and the years to follow.

Still, I expect the CCMs have a well-thought plan to send all those dead batteries to China or Romania so that they can be buried with due ceremony. Or maybe they simply plan to dump them into the sea with the rest of the recycling rubbish.

61.

Government policies mean that hundreds of local schools and rural shops and post offices are being closed. When village schools and post offices are closed, the people living in those villages must spend huge amounts of time, money and energy driving into their nearest town. When rural schools are shut, children have to be taken into town every morning and brought home again every evening. When post offices are shut, villagers must drive into town to buy a stamp, post a parcel or collect a pension. The cost in energy terms is vast. If the Government is serious about cutting our energy expenditure they would do well to make an attempt to preserve rural schools, shops and post offices.

62.

The CCMs insist that the oil in the ground (or under the sea) will have to stay where it is. They want to stop us using oil, gas, coal and uranium immediately and completely. They want us to get all our energy from renewables such as solar and wind power and through their power on social media and in the EU they are forcing through policies which fit with these aims. However, they want to carry on flying to their regular climate change conferences and using their websites and their laptops.

The CCMs want the big oil companies to go bust.

Because of the political pressure they've exerted, numerous investment companies are refusing to invest in oil companies. If they get their wish then most of us will die. The only consolation is that many of the CCMs will be among the dead. The cockroaches and the termites will have the planet pretty much to themselves.

63.

There are plans to build huge underground storage facilities where the emissions from power plants and factories could be stored. Has anyone really thought about this? I can't begin to imagine the cost and the energy required to build and maintain these massive leak-proof storage facilities? And wouldn't they be rather a prime target for terrorists?

64.

The CCMs have become very powerful and they have forced politicians, banks and insurance companies to take notice of their insane bleatings.

Potty CCMs have even pushed the Government into changing the rules in such a way that puts big pension companies under pressure to dump their oil company shares. This is lunacy and if I had a pension with a company which bowed to these loony activists I would withdraw my money immediately. There is, as usual, talk of 'stranded assets' (oil being left in the ground because of no demand) and of oil companies being unable to sell the oil they have found. The CCMs and their followers seem to think that we can all cope quite well without oil or gas and consequently without food, heating and transport.

The plan is clearly to destroy oil and coal companies, and their shareholders. It obviously has not yet occurred to the CCMs around the world that without coal and oil there won't be enough electricity for their little electric cars as well as their laptops and mobile phones.

Oh, and as I have already pointed out, you cannot make electric cars without steel. And you cannot make steel without coal. You could, I suppose, make electric cars out of plastic. But you can't make plastic without oil. And if you made electric cars out of cardboard, the EU's rules about safety would have to be slightly rewritten.

If the oil which is in the ground really does stay where it is then most of us won't care because we will be dead.

65.

People in electric or low emission cars are far less likely to stop at zebra crossings than are people driving petrol or diesel engine cars. The drivers of electric cars are, it seems, so bloody pleased with themselves that they feel they can treat everyone else with disdain and contempt.

We need a jolly giant to whom we can say: 'Please eat up our Greens.'

66.

Many national newspapers have run powerful campaigns attacking plastic bags but still put their magazines (together with the promotional inserts) into plastic bags.

We sometimes forget that plastics are vital for making many essentials in addition to plastic bags. Plastics are used for toothbrushes, window frames, spectacles, computers, furniture, mobile phones, car parts, refrigerators, contact lenses, false teeth, medical implants and currency notes. It's quite difficult to make contact lenses out of wood.

67.

The world's demand for energy is rising quite fast – thanks to developing countries which have discovered the joy of the motor car and the electric toaster and which are, rather selfishly, insisting on being allowed into the 21st century.

68.

It is important to remember that the recent enthusiasm for the ancient climate change theory (nee global warming) was created by politicians who wanted to introduce a whole raft of new laws restricting our use of energy so that there would be plenty of oil left for tanks, jet fighters and ministerial limousines.

69.

I find it alarming that Al Gore's film *An Inconvenient Truth* is still highly regarded. A British High Court Judge concluded that the film contained nine errors. If I had written a book or film which a High Court Judge ruled contained nine errors, I'd want to hide in a cupboard. But Mr Gore's video is still promoted by CCMs as if it were 'science'. Al Gore spent much time flying around the world promoting a film which explains why people who spend all their time flying around the world are destroying the planet.

When Gore found himself under fire for using 20 times as much electricity as the average American required, in his Nashville mansion, he defended himself by claiming that he offset all his carbon dioxide emissions by buying green credits.

We know all about those credits, don't we Greta?

70.

There is no doubt that climate change warnings have been grossly exaggerated. Indeed, it is now clear that the whole climate change nonsense is just another piece of pseudoscience being promoted by another well financed 'scare industry'.

Moreover, climate change conspiracy theorists make the mistake of confusing their fake science with the noble and worthy aim of caring for, and protecting, the environment. And they don't give a damn how their demands affect people.

Nevertheless, despite the evidence, the climate change myth is still so popular that anyone who publicly questions it will likely be barred from BBC programmes and university lecture halls.

(I proved in my book *The Shocking History of the EU* that the BBC was 'bought' years ago by the European Union. Today, the BBC is effectively little more than a mouthpiece for the world's most fascist organisation. These days the BBC always supports whatever potty idea comes out of Brussels.)

71.

Climate change scientists who make a good living out of the climate change myth, actively argue that climate change is the biggest threat to mankind. Formerly unknown, unrecognised and largely forgotten they have become a notable part of the modern 'Fear Game' – keep the people frightened and they won't ask too many questions. But their enthusiasm for a money-spinning apocalyptic threat all of their own is hardly surprising. The climate change myth has made them famous. All those long-haul flights to conferences in exotic parts of the world would stop if the myth was not sustained. An entire global warming industry has been built up. It's even bigger and more impressive than the now slightly shop worn AIDS industry. Tens of thousands of self-appointed experts regularly fly around the world, attend meetings, accept huge consultancy fees, set up advisory committees which meet in five star hotels in lovely beachside resorts and spend their mornings thinking up new ways to punish ordinary people and sustaining the whole myth. (The afternoons are spent on the beach and the evenings are spent having a damned good time – all at our expense.)

72.

The CCMs changed the threat from Global Warming to the considerably more 'wishy washy' Climate Change when it became abundantly clear that the world wasn't getting warmer. Then, after some bad weather, those promoting the Global Warming myth took to calling it Global Chilling. Whenever we have some sunshine they go back to calling their fake phenomenon Global Warming.

73.

Despite the threats and warnings, the vast majority of people see the climate change myth for the scam it is. Surveys have consistently shown that despite the hysteria, the lies and the propaganda most people don't believe in global warming (man-made or not).

But the CCMs still don't understand this.

On 5.2.20, the *Financial Times* newspaper (as out of touch on climate change as it was on the European Union and Brexit) suggested that 'policymakers are driven by public fears about the environment'.

No, they aren't.

Most of the public don't believe a word of the climate change nonsense they are constantly being fed.

The British rose up against the fascist EU and voted to leave. The next uprising will be against the climate change myth.

74.

Why do so many people insist on writing Climate Change instead of climate change?

I have noticed that newspapers which would use lower case initials when writing about God or the Queen seem to favour upper case initials when writing about their favourite myth. The Initial Capitals make the whole thing More Important.

75.

Although the BBC is a firm proponent of the climate change myth, it still sells its overpriced DVDs (programmes for which licence fee payers have already paid) wrapped in two layers of plastic and one of cardboard. Indeed, I rather suspect that some of the series it sells on four DVDs could be fitted onto two disks or even one disk – thereby saving even more of our planet and reducing the Corporation's carbon footprint considerably. Indeed, if they used both sides of each disk (as some companies do) the BBC could save even more material, weight and production costs. Still, we must not confuse 'saying' with 'doing', must we? And we can add hypocrisy to the long list of charges to lay against the BBC.

76.

Climate change is, in almost every respect, the 21st century equivalent to the irresponsibly exaggerated AIDS scare which scarred life in the 1980s.

The AIDS threat was deliberately blown up by homosexuals as part of a ruthless public relations campaign designed to gather attention, sympathy and support. Aggressive homosexuals spread misinformation widely – causing massive and unnecessary concern among millions of ordinary people.

One leading AIDS campaigner admitted to me that gays were deliberately distorting the facts, and frightening millions, to draw attention and sympathy to their cause. I thought it was a pretty ruthless, cold-blooded thing to do. And given that a number of people killed themselves because they were so frightened about AIDS, it was pretty murderous too. The deliberately created artificial scare about AIDS was further promoted by an unholy mish-mash of self-interested academics and business-folk.

During the 1980s, it was not considered acceptable to question the then widely held (but unsustainable) belief that AIDS was about to kill us all.

I seem to remember listening agog as the British Medical Association and the Royal College of Nursing both ignored the science and forecast, quite seriously, that we would all be affected by AIDS by the year 2000. These organisations created panic, mass fear and hysteria and were responsible for suicides. More people were killing themselves because they were frightened of AIDS than were dying because they had AIDS.

The trouble is that broadcasters love having CCMs on their programmes because hysteria produces good listening and viewing figures. Anyone trying to talk sense will be kept off the air because good sense does not do much for the ratings.

The climate change nonsense and the AIDS nonsense really are very similar. In both cases the facts have been ignored by ignoramuses who do not realise that science is, or should be, about facts and that facts demand scientific integrity and accuracy. And in both cases the promoters of the myth have a good deal to gain. The only difference is that the climate change nonsense is also being promoted by ill-informed children whereas the AIDS myth was conceived and promoted by homosexuals for their own political purposes

A few years ago, the bookshelves were groaning under the weight of books about AIDS. Today, the same shelves groan under the weight of learned rubbish about climate change. As one stupid industry dies so another comes to life.

Both myths are based on propaganda and fake science.

77.

Many CCMs ride bicycles on public roads – and take great delight in causing long tailbacks of cars and trucks. Even at the best of times, bike riders invariably produce queues of cars driving along at 10 or 15 miles an hour. At those speeds, motor vehicles are very inefficient and use far more fuel than when travelling three or four times as fast. So, bicycle riders are causing additional air pollution and doing great harm to the environment.

78.

Hospitals and factories have been asked to install back-up diesel generators because Britain now has just 0.1% of spare electricity.

And CCMs want us to stop using oil, gas and coal for generating electricity!

79.

Electric cars will never match the driving range available to cars driven by internal combustion engines because if the weight of the batteries used is increased enough to give an extended range, the extra weight adversely affects the car's performance – and the range it will travel.

80.

The UK Government has announced that car manufacturers will no longer be able to sell motor cars powered by diesel or petrol after the year 2035 – or possibly earlier. This will mean that someone will have to manufacture and install around 20 million charging points between now and then. There will also need to be a great many new power stations built. (It takes up to 20 years to build a power station so time is already running out.)

Ending the sale of petrol and diesel vehicles will mean that millions of second hand cars will have little or no residual value and will have to be destroyed.

Gas boilers and heating systems will also need to be replaced since the Government has announced that it will be illegal to install a gas boiler in a new home after 2025. As the manufacture and servicing of gas boilers and heating systems ends, so older homes will have to switch to electricity.

Disposing of all these cars and central heating systems will cost trillions and will cause severe financial hardship to millions. There will, of course, also be a massive environmental cost.

And we will need a good deal more electricity than we can possibly generate.

81

Greenpeace campaigns to have the world's oil left in the ground. But in 2014, a Greenpeace senior executive was reported to have been commuting by plane for two years. The employee, who had been flying regularly from Luxembourg to Amsterdam, was defended by the Executive Director of Greenpeace UK who seemed to think it was acceptable for the employee to fly so much so that 'he could balance his job with the needs of his family'.

In February 2020, Greenpeace bought a full page advertisement in the *Financial Times* to reprint a letter that the Executive Director of Greenpeace UK had sent to BP. The letter, demanding an end to the use of fossil fuels, was, in my view, the most arrogant letter I have ever read.

I can only assume that Greenpeace would nevertheless like BP to continue making a small amount of fuel available for planes used by its employees.

82

When Greta Thunberg went to America recently (to campaign against the use of oil, among other things) she returned on a £4 million boat allegedly equipped with two diesel engines for emergencies and a diesel fuel tank carrying 672 litres of diesel. When she was ready to return to Europe, members of the crew had to fly out to America to bring her back. She would have used less oil if she had simply climbed aboard a commercial aeroplane and flown across the Atlantic. That's hypocrisy of which Prince Charles and Harry could be proud.

If Little Greta wants to cross the Atlantic again maybe she should travel in a wooden rowing boat.

Time and time again, it becomes clear that those who shout loudest that the rest of us must not use oil or gas, seem to believe that although we must eschew those products it is perfectly acceptable for them to do so. They, it seems, are so important that the rules do not apply to them. It is this arrogance, this hypocrisy, which distinguishes all those (with a bunch of celebrities and members of the British royal family at the forefront) who shout loudest about climate change. It is this extraordinary hubris which ensures that their exhortations are ignored.

83.

In the last six years, the UK spent £900 million (and a good deal of energy) sending waste over to Europe where it was quietly burnt. Our recycling used to be sent, at great expense, to the Orient. We used to send 55% of our recycled paper and 25% of our recycled plastic to China even though carting all that stuff such long distances could hardly be described as 'green' or 'environmentally friendly'. But now the Chinese have decided they don't want our rubbish. This is no great surprise because the vast majority of the plastic we throw out cannot be recycled, or is far too expensive to recycle, and so the

Chinese couldn't do anything much with it except burn it and the fees we were paying wasn't enough for the inconvenience and the pollution.

However, we still send most of the recycling waste to other Asian countries because we don't know what else to do with it. Naturally, no one bothers to work out how much energy is wasted in transporting our waste half way round the world. Nor does anyone announce how much it costs to collect the stuff, load it onto ships, unload the ships, cart the stuff off by lorry and pay someone to deal with it at the other end.

So, bottom line, what happens now to our carefully sorted recycling material when we've finished washing it and sorting and putting it out for the recycling lorry?

What do the sanctimonious recycling gestapo do with all those trays and sacks of sorted rubbish?

Well, some of it is dumped in landfill sites, some of it is burnt and some of it is dumped in the sea.

My long-term, trenchant criticism of the whole recycling nonsense has been justified by the recent discovery that British recycling, carefully washed and sorted and collected and transported is now not even being burnt. It is currently being dumped in Poland where it is unloved, unwanted and a very expensive tribute to the stupidity of the whole recycling nonsense.

Councils everywhere send round their stupidly expensive recycling lorries, and gullible citizens spend hours washing out their empty jam jars and yoghurt cartons and collecting all their bits of uneaten food and bunging them into little plastic waste caddies. And it is all a waste of time.

It is clear that our recycling efforts are actually doing far more harm to the environment than if we just kept it at home and either burnt it or dumped it in unused mines.

Abandoning recycling and reverting to once a week bin collections would be safer, cheaper and much, much better for the environment.

But there seems little chance of that.

The Government, and the sanctimonious recycling do-gooders, have promised that Britain will continue to obey absurd EU laws – pointless and irrational though they may be.

84.
CCMs won't allow the Government to build nuclear power stations even though the pollution and risks from nuclear power are very small.

85.
The EU has forced us to replace our old, traditional, incandescent light-bulbs and replace them with compact fluorescent bulbs on the grounds that the new bulbs use less energy than the old ones. The immediate cost of throwing away our existing light-bulbs and converting our lamp holders was estimated by the British Government to be around £3 billion.

The first problem with the new EU light-bulbs is that the light produced by them is so poor that people have difficulty seeing where they are going – let alone being able to see well enough to read or work. It seems clear that the new bulbs will result in far more accidents. The second problem is that the new bulbs must be kept switched on for longer than the old ones. This means using up more energy. Third, the EU-approved bulbs are bigger and heavier than the old ones and so cost more to store and transport. Fourth, the new bulbs cost up to 20 times as much as the old ones and do not always last much longer. Fifth, the EU bulbs tend to flicker and produce a harsher, less relaxing light than the bulbs the EU has banned. It seems certain that the new bulbs will cause migraines, dizziness and possibly fits. Sixth, the new bulbs cannot be used with dimmer switches or electronically triggered security lights. Millions of pounds worth of equipment will have to be replaced. Less than half of all the light fittings in British homes will take the EU bulbs, so millions of light fittings will eventually have to be replaced. The bulbs cannot be used in ovens, freezers or microwave ovens because they don't work if the temperature is too hot or too cold. And the new bulbs cannot be used in enclosed light fittings because they need more ventilation than the old bulbs. Seventh, the new EU bulbs require ten times as much energy to manufacture as the old ones. Eighth, the new bulbs seem to exacerbate a number of health problems. For example, they can trigger eczema-like skin reactions.

Ninth, the EU bulbs need to be broken in for about 100 hours before their brightness level stabilises. Tenth, the bulbs may interfere with the remote control for your television and they may interfere with your radio or cordless telephone. Eleventh, the new bulbs don't last for long if they are turned on and off – to get the best out of them you have to leave them on permanently and learn to sleep with the light on. Twelfth, the EU-approved light bulbs use toxic materials such as mercury vapour. Apart from the fact that the bulbs could be dangerous if they are broken (the British Government has warned that if one of the EU bulbs is smashed, the room should be vacated for at least 15 minutes and great care must be taken in clearing up the debris) there is a problem with disposing of dead bulbs. The EU has banned products containing mercury vapour from landfill sites, and so used bulbs will have to be collected and disposed of separately.

To sum up, the bulbs the EU is forcing us to use (to save energy) are much more dangerous, far more expensive and less energy efficient than the old ones. Should we congratulate the light-bulb industry lobbyists for succeeding in persuading the EU to force over 500 million people to stop using a safe, efficient, cheap product and replace it with what seems to me to be a dangerous, inefficient and expensive one?

The aim was to save energy (to please the CCMs) but the new light-bulbs will require more energy than the old ones.

As an ironic footnote to this, the EU wants to ban barometers because they contain a small amount of mercury. Barometers have been around for at least 350 years and there is no record of anyone ever having been injured by the mercury inside one. The EU wants all barometers to be broken up and destroyed. The eurocrats do not seem to realise that by doing this they will release all the mercury (now safely contained) into the environment.

Despite this attack on barometers, the eurocrats do not seem to worry overmuch about the fact that hundreds of millions of individuals in the EU have been vaccinated with vaccines containing a mercury related product. Nor do they seem to worry about the fact that one of the commonest types of dental filling contains mercury.

86.

Water is in desperately short supply.

That's why pure drinking water often costs more than oil products such as gas and petrol.

Despite this simple truth, local councils in the UK insist that citizens wash out bottles, jars and tins. They are obeying absurd recycling rules introduced by the EU.

Anyone who washes out tins or bottles with fresh drinking water is wasting the planet's most valuable natural resource. And that is a crime against humanity.

(In addition, there is no little irony in the fact that local councils everywhere spend a fortune printing and distributing expensively designed and produced leaflets, booklets and newspapers describing the true Joy of Recycling.)

87.

Within five years from 2020, it will be illegal for owners of new homes to heat or cook by gas. The Government insists that we triple our planting of trees (though, bizarrely, millions of trees are being burnt as the main part of our new energy production programme) and, to please the CCMs, other schemes are being introduced which will cost taxpayers between £20 billion and £40 billion a year.

Sadly, the Government's schemes will mean that there will be a dangerous shortage of electricity for heating, cooking and industry and within a few years we will be forced to shiver and eat cold food in a nation with no viable economy.

It is, perhaps, worthwhile remembering that countries such as China and India are building massive new fossil fuel power stations. Coal seems to be the fuel of choice in many countries since it is cheap and efficient.

88.

The biggest use of plastics is now for making recycling boxes.

It would be much better for hygiene, health, the economy and the environment if all rubbish was collected in black plastic bags and used as landfill.

89.

For some time now, politicians and journalists seem to have regarded the plastic bag as the major threat to the planet.

We are told that our oceans are crammed full of plastic bags, though no one has produced any evidence for this curious claim or explained why shoppers are running down to the shore to throw their empty bags into the sea. The truth, of course, is that when plastic bags do end up in the sea it is a result of industrial dumping. And it is not the fault of the plastic bag that some litterbugs throw them around.

To ban plastic bags because irresponsible companies dump their waste into the sea, or because some of them end up littering the countryside, makes as much sense as banning newspapers because some of them end up littering the countryside.

Supermarkets and stores have run huge and sanctimonious campaigns explaining that they are banning free plastic bags in order to save the environment. In reality, of course, they stopped distributing free plastic bags in order to save themselves money and to justify the sale of so-called 'bags for life' at outrageous prices. It has not gone unnoticed that those same supermarkets still use vast quantities of paper and plastic in their packaging.

Some stores have replaced plastic bags with paper bags. Paper bags may be nice and old fashioned but they were never very practical. Try carrying one, packed with groceries, in a rainstorm. There is also the problem that you have to chop down trees to make paper bags. This damages the environment and contributes to global warming by reducing the number of trees on the planet.

Other stores are selling heavy-duty cloth bags that can be used more than once (just like plastic bags can be used more than once). These bags are usually made of cotton although making cotton bags requires a good deal of energy and a lot of water, both of which are in short supply. And in order to grow the profitable extra cotton to make the bags, farmers are growing less food. The result is that people in Africa are starving to death so that nice, liberal do-gooders can wander around carrying their shopping in cotton bags.

The hypocrisy seems unending.

Newspaper owners and broadcasters have run excited campaigns telling us all that if we use plastic bags we are threatening our children's future.

How true is all this?

Well, actually, it's not very true at all.

Plastic bags are manufactured from a form of plastic derived from a by-product of the oil industry. If the plastic bags weren't made, the by-product would have to be burned off – increasing carbon dioxide emissions. So people who use plastic bags are actually helping to save the planet – not destroy it. And, the self-righteous CCMs who campaign against plastic bags are actually campaigning against the planet's best interests. The manufacture of plastic bags consumes less energy than the manufacture of paper bags. Plastic bags generate less solid waste than paper bags. And they are responsible for fewer atmospheric emissions.

In reality, there is a lot to be said in favour of plastic bags. They are widely reused. Few things are recycled as often or as efficiently or as extensively as plastic bags. It is more difficult to reuse paper bags and most modern plastic bags are biodegradable.

It has been claimed that plastic bags are responsible for killing millions of birds every year. And that they kill 100,000 marine animals too. In fact, there is no evidence to support these claims. The original source for this propaganda was a report published in 1987 which found that between 1981 and 1984, around 100,000 animals had been killed by discarded plastic fishing nets. This report was subsequently misquoted and the plastic fishing nets became plastic bags. There is not, and never was, any evidence showing that plastic bags kill millions of birds every year. (Just how the bags were supposed to kill the birds was never explained.)

90.

There isn't a material on the planet which is loathed quite as much as plastic. I have even heard CCMs claim that the small plastic toys which were included in breakfast cereal packets in the 1960s have somehow become a major threat to our environment.

It isn't difficult to conclude that the loathing has been overdone.

The amazing thing is that politicians insist on ignoring their own experts.

In 2005, the Scottish Government reported that the manufacture of paper bags consumes four times more water than the manufacture of plastic bags. (Water is, of course, one of the most valuable and scarcest commodities on the planet.)

And a 2011 UK Environmental Agency study found that paper bags contribute three times more to global warming than plastic bags. The change from plastic bags to paper bags will require seven times as many trucks to deliver the bags. All that diesel and all that pollution! The manufacture of paper bags requires more energy and water than making plastic bags.

Moreover, cotton bags (the type loved so dearly by so-called environmentalists) are also far worse than plastic bags. Denmarks's Ministry of Environment and Food found that when all things are taken into consideration, cotton bags are far more damaging than plastic bags. A cotton bag needs to be used 7,000 times to be as good for the environment as a plastic bag. And a bag made with organic cotton needs to be used 20,000 times to be as good as a plastic bag.

Despite this evidence, when Mrs Theresa May was Prime Minister she announced that plastic was 'one of the greatest environmental scourges of our time'. Her Government has continued the demonization of plastic, and today every sanctimonious body in the country is busy telling the world that they are saving the planet by banning plastic straws.

The real reason for this nonsense?

The oil is running out but governments don't like to tell us that.

91.

CCMs demand that driving speeds should be kept low to protect the environment.

As always they are wrong.

Diesel engines are most efficient when working at 80% of their potential, and petrol engines are most efficient when working at between 60% and 75% of their potential. What this means in practice is that a petrol driven motor car with a top speed of 120 mph will be most efficient (and kindest to the environment) if travelling at between 72 mph and 90mph. If environmentalists really want us to use less fuel they should encourage motorists to drive faster, not slower. And the police should be ordered to arrest drivers who are travelling slowly.

92.

In 2019, laws came into force requiring pension scheme trustees to prepare policies for protecting pensioners against financial risks arising from climate change and other environmental, social or governance issues (these are now known as ESG issues).

When some pension scheme trustees decided that this was nonsense, the official Pensions Regulator was urged to take action and to ensure that trustees did as they were told – whether they agreed with the premise or not.

The then Pensions Minister stated that trustees 'had responsibilities and legal obligations to protect people's pension pots from climate change risks'.

If the Government really wants to protect pensions they would be better advised to stop CCMs from exaggerating the risk of man-made climate change and damaging the value of investments held by pension funds.

In 2019, European investors poured around £100 billion into so-called 'sustainable' funds which avoid investing in companies involved in anything related to fossil fuels. The theory was based on the flimsy theory that oil, coal and other materials will be 'stranded' assets as we come to rely entirely on renewable forms of energy. These ill-informed investors and advisors claim that $900 billion worth of oil and other fossil fuels will be 'stranded' and unusable.

Since it's quite clear that we will not be able to avoid reliance on oil, coal and so on for generations to come, this theory is half-baked at best.

Those who insist that coal and oil are going to become stranded assets might like to know that China and India are planning to build 800 new coal fired power stations.

93.

I do not believe that we can trust what we are told about global warming or any other scientific 'truth'. What we are fed are facts that are convenient to the people with the power, the money and the ability to use legalised violence (that, of course, is what governments are).

Governments around the world keep signing up to ever more ambitious plans to deal with climate change. None of their promises will be kept, of course. None of their targets will be met.

But it doesn't really matter. And they know it doesn't matter.

The energy shortage (and increased prices) created by peak oil will ensure that people will use less and less energy and waste less and less, thereby reducing the problem.

Unless they are stupid beyond even my wildest despair, politicians must realise this, of course.

The almost hysterical attempts to warn us of the dangers of global warming, and to persuade us to use less oil and other fuels, are, in reality, merely their way of conserving fuel sources which are disappearing and of getting us accustomed to life without oil.

Why don't they just tell us the truth – that the oil is running out and we have to use it more sparingly?

Well, that might cause panic and it would certainly enable more people to see just why we are still fighting wars in countries which have rich oil deposits.

Moreover, politicians find lies much easier to tell than the truth. And the war on global warming has much in common with the war on terrorism.

94.

Britain's electricity generating system is about to collapse.

In order to please the EU, British governments have closed down old-fashioned, traditional coal fired power stations.

But the closed power stations haven't been replaced with anything useful.

Where do these folk who drive electric cars think the electricity comes from?

Britain has had to introduce special diesel powered generators to provide enough electricity to keep electric cars on the move.

Every motorist who buys an electric car is making things worse and the chances are that electricity will soon have to be rationed with the result that shops, factories and homes will have to go back to the three day week.

Energy prices will rise rapidly.

The cuts in electricity supply will endanger the health of millions.

And it will be the CCMs and the electric car owners who must share the blame.

95.

There are still some CCMs who eat meat.

It cannot be said too often that this is remarkable for two reasons.

First, the increasing demand for meat and meat products is one of the reasons for the world food shortage.

Second, raising animals for food generates more greenhouse gases than all the cars and trucks in the world combined. Anyone (including celebrities and royals) who professes to be concerned about global warming but does not also campaign for vegetarianism is, I fear, something of a hypocrite.

96.

The global warming doomsters want us to cut out carbon emissions by 80% from the 1990 level. The only way we will hit this target is by cutting our Gross Domestic Product by 80% too. That will cost every British household £2,200,000 between now and 2050. And it will mean no central heating, no cars, no television, no computers and no washing machines.

97.

Leading climate change campaigners (a disparate bunch of pop singers, royals and politicians) are making names for themselves by flying around the world in private aeroplanes instructing the rest of us that we must stay at home because flying around the world is damaging the planet.

The celebrities involved are usually the same group of multimillionaires who, having made fortunes out of flogging overpriced albums, concert tickets and T-shirts, own several large, energy consuming homes. They often own fleets of limousines and live as tax exiles. They attempt to ingratiate themselves with their fans by making critical remarks about consumerism and capitalism.

Now, I don't know about you but I don't want to take any advice (or be hectored) on any aspect of the environment (or anything else) by self-indulgent divas who have their own jets and employ a retinue of servants. Such people have no right to preach to the rest of the world about poverty, global warming or anything else, without attracting criticism that they are doing it for the publicity.

98.

Politicians are trying to persuade people other than celebrities and members of the royal family to fly less. But at the same time they are thinking of allowing Heathrow airport to have a third runway. It's like telling people they can't smoke and then building another cigarette factory.

99.

The output of existing oilfields tends to fall by between five and seven per cent a year. That is an unfortunate fact of life that not even Little Greta can change. This is a problem because the world is going to continue to need oil for decades to come. Even if the UK stops using oil, there is little chance that China, India and the United States (for example) will stop their oil consumption. So the conclusion has to be that the world will need a good many new oilfields to continue supplying oil and these will have to be found on a planet where the supply of oil is running out.

But the CCMs are forcing banks to refuse finance for oil companies.

So, disaster lies ahead.

100.
I wonder if the drivers of electric cars (who currently pay no road tax and who enjoy large subsidies paid for by taxpayers) realise that when petrol and diesel cars are removed from the roads, the Government will need to raise money to pay for the building and maintenance of roads (and for other more general expenditure).

At the moment, fuel duty brings in over £28 billion a year. There will need to be a tax on electric cars and given the cost of the infrastructure required (in addition to roads) the cost will probably be between £1.50 and £2.00 per mile travelled. So a motorist in an electric car will need to pay around £30 to £40 a day for a round trip commute of 20 miles. A trip of 200 miles will require a payment of £300 to £400.

101.
If the hysterical CCMs could put down their placards long enough to do a little research they would find that Britain's use of energy fell by a fifth between 2000 and 2020. We travel less and our cars go further on a gallon of petrol. Our air is cleaner than it has ever been. We have taken the carbon out of economy faster than any other country. Through caring, through the activities of capitalists and through technology we have made enormous strides – without the slogan shouting vandals and terrorists demanding that we listen to their nonsense.

Nevertheless, the British Government is now in thrall to these lunatics and is introducing unilateral plans which will do dramatic damage to our way of life.

Maddest of all, the Government is forcing drivers to use electric cars which are known to be more damaging to the planet than diesel and petrol driven cars. And those electric cars will require vast quantities of electricity which will have to be produced by burning oil or wood. (Calling wood 'biomass' doesn't change what it is.)

The myth of man-made global warming is a cruel marketing strategy designed to introduce us to the reality of peak oil without actually bothering to tell us that the oil is running out.

The celebrities, royals, CCMs and schoolchildren are merely being used to help sell the myth.

Afterword
The Harm the CCMs Are Doing

There is no reliable scientific evidence that our climate is changing. There is no reliable scientific evidence that 'climate change' is man-made. And there is no evidence that 'climate change' is going to destroy the planet.

There is evidence, however, that the pious, sanctimonious terrorists who are deliberately trying to terrify us all with their half-baked nonsense are causing enormous distress.

The self-indulgent hysterics who insist that we have only so many years, months or weeks to save the planet have terrified millions quite unnecessarily. This is intellectual terrorism. Climate change is a mean myth fostered by self-serving pseudo scientists and promoted by ignorant celebrities whose stupidity is matched only by their arrogance and their staggering hypocrisy.

The irony is that the utterly daft regulations, policies and taxes being introduced as a result of powerful lobbying by the CCMs will do far more harm than any changes in our climate.

The CCMs who push for extremist policies in the name of climate change are doing massive harm.

Just as the nutters who attempted to defy the result of the Brexit referendum caused uncertainty, confusion, economic loss and infinitely more short-term damage than leaving the EU without a deal could possibly have done, so the CCMs (to a large extent the same individuals) are causing massive harm to the nation and the planet. It is downright silly to assume that solar energy and wind power can ever replace fossil fuels.

The climate change con is the biggest piece of mass chicanery and deceit since the AIDS myth was promoted with such vigour, and it seems likely to be longer lasting and more damaging.

It is not climate change (aka global warming aka global cooling) which will destroy the world's economy, and lead to millions of deaths, but the hysterical threats, demands and campaigns organised and coordinated by the most hysterical of the CCMs.

Politicians and business folk who are desperate to appease the ill-informed CCMs are going to wreck the planet by agreeing to their daft demands to stop using fossil fuels.

The coming problem is not going to be increasing temperatures – but an absence of energy. Our madnesses are going to lead to power outages that will last for days or weeks rather than just minutes and hours.

By closing down electricity plants which rely on coal and gas and by not building alternative power plants (such as those powered by nuclear power) we are creating a massive problem. The British Government's attempts to lead the world in appeasing the CCMs, and obeying daft EU laws, mean that Britain is now probably the most vulnerable nation on the planet. Millions are going to freeze or starve to death because of these insane policies.

I sometimes wonder if the supporters of terrorist organisations which preach about climate change are keen to see the human race become extinct. (The CCMs don't like being called terrorists but the definition of terrorism is 'the unofficial use of violence and intimidation in the pursuit of political aims'.)

Are these sanctimonious prigs knowingly working for the establishment (and helping to force through the changes which are necessary as the oil runs out) or are they just doing so unwittingly?

The climate change mythmakers ignore the considerable downside to their hysterical demands. Millions of people have suffered (and are still suffering) because of the ill-based exhortations of the climate change mythmakers. Poor people, and those in developing countries, are the ones who are suffering the most. There is no doubt that the hysterical climate change advocates are doing far more harm than good, and that the only danger to life and to society comes entirely from those who insist that man-made changes to our climate demand massive, immensely harmful changes to our way of life.

CCMs are threatening to close down London for weeks at a time. This will cause inconvenience but it will also cause much distress, real hardship and a good number of deaths. But, amazingly, these self-righteous, simpering, ingloriously ignorant vandals demand, take and are given the moral high ground. The police seem to have decided to allow these lunatics to close down towns and cities with impunity. Maybe that's what the politicians have told them to do.

Today, the climate change hysterics claim (with no evidence whatsoever) that our planet will be unliveable within a generation. They are promoting a myth and they are wrong for a number of reasons. It is downright silly to assume that solar energy and wind power can ever replace fossil fuels. If our politicians continue on their reckless path to rely solely on energy obtained from renewables then most of us will either freeze to death or starve to death.

It is, I fear, too late to persuade the CCMs that they are wrong. Their arrogance will not allow them to retreat from the position they have established. I would like to think that Greta Thunberg, and those who share her views, will spend a little more time studying and understanding the facts. But I suspect that won't happen.

It is up to the rest of us to promote the view of science and sanity and to persuade politicians and opinion makers to listen to the truth.

If you found this book of interest we would be very grateful if you would write a review on Amazon. The CCMs will doubtless do their best to smother the book with bad reviews – even though they haven't read it – so your help will be vital.

Printed in Great Britain
by Amazon